# THE New Standards AND Accountability

## Will Rewards and Sanctions Motivate America's Schools to Peak Performance?

by
Larry Lashway

ERIC
Educational Resources Information Center

ERIC Clearinghouse on Educational Management
College of Education
University of Oregon
2001

**The New Standards and Accountability: Will Rewards and Sanctions Motivate America's Schools to Peak Performance?**

Editorial Director: Stuart C. Smith

Graphic Designer: Leeann August

---

**Library of Congress Cataloging-in-Publication Data**

Lashway, Larry.
The new standards and accountability : will rewards and sanctions motivate America's schools to peak performance? / by Larry Lashway. — 1st ed.
p. cm.
Includes bibliographical references and index.
ISBN 0-86552-149-2
1. Educational accountability—United States. 2. Education—Standards—United States. 3. Motivation (Psychology) I. Title.

LB2806.22.L37 2001
379.1'58'0973—dc21

2001033182

---

Type: 11/13 Palatino

Printer: McNaughton & Gunn, Inc., Saline, Michigan

FIRST EDITION
Printed in the United States of America, 2001

THE ASSOCIATION OF EDUCATIONAL PUBLISHERS Ed Press

ERIC Clearinghouse on Educational Management
5207 University of Oregon
Eugene, OR 97403-5207
Telephone: (541) 346-5044 Fax: (541) 346-2334
World Wide Web: http://eric.uoregon.edu

ERIC/CEM Accession Number: EA 030 821

This publication was prepared in part with funding from the Office of Educational Research and Improvement, U.S. Department of Education, under contract no. ED-99-C0-0011. The opinions expressed in this report do not necessarily reflect the positions or policies of the Department of Education. No federal funds were used in the printing of this publication.

## MISSION OF ERIC AND THE CLEARINGHOUSE

The Educational Resources Information Center (ERIC) is a national information system operated by the U.S. Department of Education. ERIC serves the educational community by disseminating research results and other resource information that can be used in developing more effective educational programs.

The ERIC Clearinghouse on Educational Management, one of several such units in the system, was established at the University of Oregon in 1966. The Clearinghouse and its companion units process research reports and journal articles for announcement in ERIC's index and abstract bulletins.

Research reports are announced in *Resources in Education (RIE)*, available in many libraries and by subscription from the United States Government Printing Office, Washington, D.C. 20402-9371.

Most of the documents listed in *RIE* can be purchased through the ERIC Document Reproduction Service, operated by Cincinnati Bell Information Systems.

Journal articles are announced in *Current Index to Journals in Education. CIJE* is also available in many libraries and can be ordered from Oryx Press, 4041 North Central Avenue at Indian School, Suite 700, Phoenix, Arizona 85012. Semiannual cumulations can be ordered separately.

Besides processing documents and journal articles, the Clearinghouse prepares bibliographies, literature reviews, monographs, and other interpretive research studies on topics in its educational area.

## ADMINISTRATIVE STAFF

**Philip K. Piele,** Professor and Director
**Stuart C. Smith,** Associate Director

## CLEARINGHOUSE NATIONAL ADVISORY BOARD

# Preface

When Stuart Smith at the ERIC Clearinghouse on Educational Management offered me the opportunity to write a monograph on accountability, I accepted without hesitation. The topic was both timely and substantive.

Moreover, as a teacher and administrator myself, I have a strong professional interest in the issue. Most of the time, I have reasonable confidence that I'm earning my salary, but if someone asked me to prove it, I'm not sure I could come up with a convincing answer. In a profession devoted to the long-term growth of hearts and minds, what could count as measurable results?

In retrospect, a little hesitation would have been in order. Accountability has become the 800-pound gorilla of school reform, and it casts its shadow on just about every policy debate in education. Every time I read a book or article, I seemed to find a new dimension to the issue, and my research would merrily head off in an entirely different direction.

Fortunately, early in the process I discovered the Southern Educational Research Board's monograph on accountability, with its five-part analysis of standards, assessment, public reporting, incentives, and professional development. That struck me as a logical structure, and it allowed me to keep the project manageable and reasonably coherent.

As always, my intent throughout the process has been to provide a concise overview of the issue for school leaders, teachers, policymakers, and whomever else might have a professional interest in the topic. While that audience is always looking for practical, go-right-out-and-try-it solutions, in this case they may have to settle for a better understanding of the issues. The accountability literature is still short on recipes. At best, a diligent searcher can find clues, but those clues will be useful only with reflection, sensitivity, and no small amount of courage. My hope is that conceptual clarity will at least provide a starting point for the leaders who have to guide their schools through the accountability maze.

I do, however, urge readers to exercise caution in the conclusions they draw. First, although I have rather loosely used the term *research literature* to describe the sources for this work, most of what is out there is not the kind of rigorous experimentation we associate with that term. Certainly, there is nothing in the research that shows that accountability works—or doesn't work. It is simply too early for that kind of conclusion. Instead, most writings on accountability range from thoughtful analysis to ardent advocacy. The best are very, very good, but still subject to further debate.

Second, whatever their source, the ideas presented in this book have gone through the same filter: me. While I try to maintain reasonable objectivity and openness, my efforts to synthesize such a large body of work are bound to have biases and blind spots. For that reason, when readers encounter an idea they want to pursue, I encourage them to use it just as a starting point, not as the definitive word.

The nature of this kind of research has changed since I did my first ERIC/CEM monograph several years ago. The range and richness of Internet-based resources grows steadily, and like many other researchers, I have been gratified by the quality and the convenience of those materials. However, we have not yet reached the point where libraries are dispensable, and I continue to be blessed with an abundance of good ones: the Washington State Library, the Olympia Timberland Library, and the libraries of The Evergreen State College and Pacific Lutheran University.

As always, I'm grateful to the Clearinghouse for offering me the professional challenge of undertaking this work, and especially to Stuart, who was supremely supportive throughout a lengthy process that began in one millennium and ended in another. Stu's ability to see past the ever-slipping deadlines to the finished product played no small part in keeping me on track.

Larry Lashway

# Contents

# 1

# The Accountability Challenge

*While the accountability movement pushes forward with considerable fanfare, behind the scenes multiple perspectives and conflicting agendas are clashing, with uncertain results.*

In November 1849, Levi Hubbard, clerk of school district number 5, Town of Sheboygan Falls (Wisconsin), noted in his record book:

> Miss Margaret Ross commenced teaching the summer school & taught five weeks at $1.25 per week. She did not give general satisfaction & left the school.

School boards in 1849 didn't demand much of their teachers—good moral character, an orderly schoolhouse, slightly more knowledge than their students—but when they were disappointed, they didn't hesitate to act. Accountability was a simple matter: If the board judged that a teacher was not living up to expectations, it dismissed her.

This sort of accountability was viewed as purely a matter of individual responsibility, and if there was local gossip about the teacher's lapse, there was no great public outcry. School #5 found Margaret Ross's performance unsatisfactory and took what it considered appropriate action; the board would not have considered the possibility that schools or teachers in general were failing to live up to their responsibility to society.

Today, by contrast, educational accountability is at the center of a national debate that has raised fundamental questions about school effectiveness, institutional reform, and human motivation. State legislatures around the country are revamping educational policies, spending millions on new assessment systems, and redesigning compensation systems in an effort to ensure that schools deliver the promised results. Consider these news items:

- In Denver, teachers have agreed to pilot "pay for performance" salary schedules that link teacher compensation to student learning (Denver Public Schools).
- In Virginia, the state's new "Standards of Learning" are so rigorous that only 6.5 percent of public schools met the benchmarks on the second round of testing (Jessica Portner 1999).
- In Texas, schools whose students too often fail the licensing exam can lose their accreditation ("Accountability System for Educator Preparation").
- In Florida, students attending schools that flunk the state's report card (two Fs in four years) receive vouchers worth $4,000 to be used at any public or private school of their choice (Jessica Sandham).
- In Jefferson County, Colorado, school district officials are linking their request for additional tax funds to a promise to improve student performance. For a $25 million annual increase, the district will guarantee a 25 percent improvement in reading and math scores in two years (June Kronholtz 1999).

These examples—unimaginable just a decade ago—suggest that policymakers have at last gotten serious about accountability, and that school leaders face a changed landscape of public expectations. Whereas it was once sufficient to run an orderly, well-organized school with qualified teachers and up-to-date curriculum, it now appears as though *results* will be the universal yardstick of leadership.

A closer look, however, reveals a more ambiguous picture. Whereas all states have mandated testing, and forty-eight publish school or district "report cards," only eighteen link graduation to test results, only twenty offer schools monetary incentives for good performance, and just one ties teacher evaluation to student performance (Ulrich Boser). Even as they try to wrestle their schools into compliance with the new policies, leaders may find it hard to stifle some nagging doubts. Are policymakers serious about accountability? Will they stay the course when the hard decisions have to be made?

Moreover, school leaders serve more than one master, and "accountability" carries different meanings for different stakeholders. When directly asked, "Should schools be more accountable?" virtually everyone reflexively responds, "Of course," but their visions may differ.

For policymakers, accountability usually seems to mean that students perform well on tests. For teachers, it is often defined as working hard to meet the needs of students. For parents, it may mean simply that their wishes are listened to. For some educational critics, it means that if students are going to be held to higher standards, the public should provide the resources that give all students an equal opportunity to learn. Thus, while the accountability movement pushes forward with considerable fanfare, behind the scenes multiple perspectives and conflicting agendas are clashing, with uncertain results.

## The Many Faces of Accountability

At the most basic level, accountability is an "accounting," a way of explaining one's actions to those who have a right to the explanation (Robert Wagner 1989). The prototype is the Biblical parable of the talents, in which three servants are entrusted with the care of their master's money. Upon the return of the master, each is called upon to explain what he has done with the money. The two who invested the money and earned a profit are praised and amply rewarded. The third, knowing the master to be a hard man, has declined to take any risks and simply buried the money. For this he is soundly condemned.

What is not clear in the parable is whether the third servant is being chastised because of his play-it-safe philosophy or because he failed to turn a profit. In the same way, policymakers have often wavered between two definitions of accountability. In the traditional "input model," educators are considered responsible for following professional standards and best practices. For example, most accreditation protocols examine whether schools follow generally accepted practices; if they hire qualified teachers, provide adequate libraries, and have a well-defined curriculum, they gain approval. The more recent "output model" pays little attention to inputs and demands results; educators are promised autonomy in their practice but in return must deliver improved student achievement.

In addition to these contrasting perspectives, schools operate in a complex governance system in which accountability is demanded at different times in different ways by different participants. Linda Darling-Hammond (1989) has identified at least five variants: political, legal, bureaucratic, professional, and market accountability.

**Political Accountability**

Schools are public institutions in a democratic society, required to subject policy decisions to public scrutiny and electoral discipline. Legislators and board members, knowing that their continued tenure depends on public satisfaction, have an incentive to respond to the wishes of voters. Much of the current push for accountability has been generated by political processes.

However, in a representative system, political accountability is a blunt instrument at best. Voters can only pass judgment on the politician's record as a whole, and a vote for a candidate cannot be easily attributed to his or her stand on a particular issue. Between elections, policymakers have to rely on polling, informal discussions, and the daily mail to determine how voters feel about any issue.

In addition, political accountability is based on constituents' *wishes*, which are not always directly linked to demands for student achievement or professional standards. For example, despite their desire for higher standards, parents sometimes resist efforts to increase the amount of homework their children are assigned (Romesh Ratnesar). Voter opinion, even when strongly expressed, does not always add up to a coherent program.

### Legal Accountability

Schools have certain legal obligations, and citizens with complaints can petition courts to intercede on their behalf. Issues involving equal opportunity, discrimination, or special education have increasingly been resolved in this way. However, most courts have shied away from adjudicating instructional accountability. For example, lawsuits built around "educational malpractice" (such as failure to teach a student to read) have been received coldly by judges reluctant to substitute their judgment for that of professional educators. For the most part, legal accountability merely establishes a framework of acceptable practices within which schools must work.

### Bureaucratic Accountability

Because policymakers tend to be remote from classrooms, most public institutions have turned to some form of bureaucracy to ensure that desirable practices are followed. State and district offices translate the wishes of legislators into policy and issue rules that educators are expected to follow. This has been the most visible form of educational accountability; a look at any state's administrative code will turn up dozens or even hundreds of specific policies to which schools are held accountable. The combination of rule setting and enforcement provides assurance to the public that schools are operating in acceptable ways.

Bureaucratic accountability assumes, however, that students will uniformly benefit from the establishment and enforcement of standardized one-size-fits-all policies—an idea that seems increasingly out of tune with today's diverse school populations. Moreover, traditional bureaucratic accountability focuses on practices rather than products. Teachers who submit a lesson plan of the required form have done their duty according to the system—even if students fail to learn.

### Professional Accountability

Professional accountability focuses on practices that are client-oriented and knowledge-based. It assumes that certain "best practices" (learnable from research and reflective teaching) form the basis of professional responsibility. But because students are so diverse, these practices must be modified and attuned to the needs of individual students. Professional accountability thus operates on the local level, where individual practitioners determine what they owe students, practice their craft, and then judge the results.

For this type of accountability to work, teachers must be well trained, knowledgeable, and dedicated to students; if not, their practice will be idiosyncratic rather than consistent with professional principles. Most important, professional accountability focuses on the teacher's responsibility to students rather than to policymakers. Adherence to standardized bureaucratic rules frequently clashes with responsibility to individual students.

### Market Accountability

In the view of many school critics, the highest form of accountability is "market discipline"—the need for vendors in a competitive marketplace to satisfy customer demands. In the market model, a school would give parents what they wanted, or the parents would take their business elsewhere, leaving the school to wither away.

As conservative critics often observe, market accountability has played a limited role in American education because public schools, for all practical purposes, have held a monopolistic position. Anyone who wishes to offer alternative instructional strategies or content can set up a private school, but without the financial support offered to public institutions; parents seeking an alternative for their children can go outside the system, but at considerable personal expense.

Within the past decade, however, charter schools, voucher plans, and other forms of educational choice have taken root across the country, and market thinking is playing a strong role in current accountability debates. In an era of skepticism about government, the idea that schools should compete "like everyone else" is intellectually appealing and emotionally satisfying.

As business executives Lou Gerstner and colleagues (1994) put it, "Results are not achieved by bureaucratic regulation. They are achieved by meeting customer requirements, by rewards for success and penalties for failure. Market discipline is the key, the ultimate form of accountability."

Despite its strong influence on current rhetoric, the marketplace still plays an ambiguous role in the accountability system. In a pure market model, there would be no role for state-mandated standards or for bureaucratic punishments and rewards; parents would simply select the schools that best satisfied their own standards of quality. There are few signs, however, that the public is prepared for such a strong market-based strategy.

While endorsing the idea that competition is healthy, the public seems to be uninformed and disengaged on educational choice (Steve Farkas and colleagues 1999). Thus far, choice mechanisms have operated on the fringes, in urban environments where performance is lowest and public frustration is highest, and where the attitude seems to be, "What have we got to lose?"

In most cases, market strategies seem to be designed as object lessons (showing how scrappy little charter schools can get the job done) or as threats (as in Florida, where vouchers will go to parents in failing public schools). Recently the Education Commission of the States has recommended an "evolutionary" two-tier governance system that would allow public school districts to directly operate schools as they have traditionally done (although with careful attention to results) or to operate as overseers of independently operated (but publicly funded) schools. The commission believes this approach will "preserve public education and build on strengths of the prevailing system, and... infuse it with a greater capacity for adaptability, flexibility, and accountability" (Education Commission of the States 1999).

## Finding a Coherent System

These multiple layers of accountability make it difficult for policymakers and educators to establish a focused, coherent strategy. The kind of accountability demanded by one constituency may be unrelated to, or even inconsistent with, the demands of another. For instance, Charles Abelmann and Richard Elmore (1999) found that many schools had internal professional standards of accountability that did not match externally imposed bureaucratic demands. When the external expectations conflicted with internal standards, teachers tended to ignore or co-opt the outside standards.

Despite this complexity, policymakers in recent years have converged on a combination of strategies that, for all practical purposes, constitute a uniform model of accountability. Particularly at the state level, policymakers have made impressive progress in forging a unitary system that incorporates an interlocking set of practices and structures designed to collectively facilitate student achievement. Specifically, states have moved toward a system that sets clear goals (standards), assesses student progress, reports results to the public, incorporates incentives and sanctions based on the results, and provides resources for carefully targeted teacher-development activities (Southern Regional Education Board 1998).

## Standards

Standards—clear statements of academic expectations—are the heart of the new system. The American tradition of local control has resulted in a system in which each district (sometimes each school) defines the learning goals for its students. On the surface, schools across the country seem to share considerable similarity in purpose; from one district to the next, there is not much difference in curricula, textbooks, and philosophy. But a closer look tells a different story. Mike Schmoker and Robert J. Marzano (1999) note:

> There are enormous differences in what teachers teach in the same subject and the same grade level in the same school. Even when common, highly structured textbooks are used as the basis for a curriculum, teachers make independent and idiosyncratic decisions regarding what should be emphasized, what should be added and what should be deleted.

They add that the perception of a common, coherent program is a "delusion."

In recent years, however, policymakers have made a concerted effort to define *learning expectations* not just as broad goals but as concrete objectives with specific benchmarks at different grade levels. In Washington State, for example, legislators identified four "essential learnings":

- GOAL I: Read with comprehension, write with skill, and communicate effectively and responsibly in a variety of ways and settings.
- GOAL II: Know and apply the core concepts and principles of mathematics; social, physical, and life sciences; civics and history; geography; arts; and health and fitness.
- GOAL III: Think analytically, logically, and creatively, and integrate experience and knowledge to form reasoned judgments and solve problems.
- GOAL IV: Understand the importance of work and how performance, effort, and decisions directly affect career and educational opportunities.

With these broad goals in place, the state then developed progressively more specific outcomes for each goal. In Goal II, for example, the mathematical component is broken down into five outcomes:

1. The student understands the basic concepts and procedures of mathematics, how to use them, why they work.
2. The student uses mathematics to define and solve problems.
3. The student uses mathematical reasoning.
4. The student effectively communicates mathematical ideas in both everyday and mathematical language.
5. The student understands how mathematical ideas connect to other subject areas, real-life situations, and career goals.

Each of the above as subgoals is then further reduced to even more specific outcomes. For example, to demonstrate mastery of subgoal 1, students must be able to understand and apply number sense, measurement, spatial sense, probability and statistics, and functions and relationships. Finally, the state provides benchmarks at several grade levels. By fourth grade, for example, students should be able to "use objects, pictures, or symbols to demonstrate understanding of whole and fractional numbers, place value in whole numbers, and properties of the whole number system." Seventh-graders should be able to "use pictures and symbols to demonstrate understanding of fractions, decimals, percents, place value in non-negative decimals, and properties of the rational number system."

The psychology behind the standards model is simple: people tend to live up to expectations. Jean Johnson and Ann Duffett (1999) put it this way:

> Central to the public's belief in higher standards is what amounts to a philosophical rule of thumb for dealing with children. Ask more from them, and they will do more. Ask less, and they will do just enough to get by. This belief is especially powerful for many people because it often stems from experiences they have had in their own lives. In focus groups, participants often tell stories about teachers, parents, bosses or even drill sergeants who challenged them, and as a result, brought them to a higher plane of accomplishment. If it's been true in my own life, people reason, then it will work for others as well.

In addition, standards act as a compass, helping teachers decide what content and activities are of most worth. Schmoker and Marzano note, "A well-articulated focus unleashes individual and collective energy. And a common focus clarifies understanding, accelerates communication, and promotes persistence and collective purpose." For both students and teachers, knowing exactly what they are expected to do makes it more likely they will mobilize their energies to meet those expectations.

**Assessment**

Testing students has long been a routine activity in schools, but much of this assessment has been only loosely tied to curricular goals. Educators have tended to view test results as a measure of individual student achievement rather than of collective school effectiveness. Most standardized tests are generic instruments that sample broad subject domains and therefore offer limited feedback on the quality of content being taught in particular classrooms.

Thus many states are developing their own criterion-referenced assessments that are closely linked to state learning standards. In this way, policymakers can get an annual snapshot of student progress on state-mandated goals, and schools can get clear feedback on their effectiveness in helping students achieve those goals.

Statewide assessments are also attention-getters. Students typically keep a close eye on assessments; when teachers warn that "this is going to be on the test," students sit up, tune in, and begin highlighting. For teachers as well, incorporating certain material on a statewide assessment carries the message, "This is important." In examining the way that Washington schools responded to new state assessments, Robin Lake and colleagues (1999) noted that, whatever else one could say about the tests, educators did take note and respond to them.

Thus, while accountability involves much more than a test, well-focused, high-profile assessments lend credibility and moral authority to the standards they reflect.

### Reporting Results

Test results are effective motivators only if communicated to those who are in a position to take corrective action: teachers, students, parents, policymakers, and the public at large. The new accountability emphasizes the need to inform the public, and most states now require some form of "public report card."

The report cards are not limited to test scores, but incorporate a wide variety of information considered useful to stakeholders. In fact, no two states have identical report cards (Lynn Olson 1999b). The data may include such things as attendance rates, school safety, dropout rates, teacher qualifications, promotion rates, teacher salaries, and class size. The reports may be lengthy, providing interpretation as well as raw data, or they may simply present numbers without much explanation.

The theory behind public reporting is simple. When stakeholders know how schools are doing, their decisions are better informed. When schools are identified as low performers, their constituencies will place pressure on them to improve; when schools are identified as high performers, they will attract new students or serve as models for other schools. When taxpayers see positive results, they will feel their support of the schools is well spent.

### Consequences

Perhaps the most compelling feature of the new accountability is the notion that performance should have tangible consequences. Good performers should be rewarded; marginal performers should face sanctions.

Traditionally, public education has measured accountability in terms of inputs. Teachers who are appropriately qualified, who show up for work, and who follow accepted professional standards in their teaching can expect continued employment and full compensation, no matter how their students perform. Other than their own inherent sense of responsibility, teachers have little incentive to make sure that students reach the goals.

Fred Newmann and colleagues (1996) described a typical scenario:

> Teachers at Fremont High gave state mandated achievement and basic skills tests and district criterion tests in the academic subjects. But no one at the school, district or state seemed to do much with the results. Scores were not published and there were no formal consequences for either the school or individual teachers tied to results. Most staff reported that they felt little or no pressure for student success on the tests.

In such a setting, when students fail to learn, teachers may agonize over the lost opportunities, mildly regret that things didn't go better, or simply look forward to the summer. But life goes on, and next year a new group will show up at the classroom door.

Much the same can be said of students. Some, motivated by self or family to gain entry to elite colleges, work exceedingly hard; those who are not so driven find few compelling reasons to exert themselves. Many freely admit they and their peers would achieve more if pressed harder (Ann Bradley 1997). Schools give grades, of course, but grade inflation has made it easier for students to achieve respectable marks with minimal effort.

For all the emphasis on standardized achievement tests, students suffer few consequences for poor scores. Frank, a college student in Washington State, recalls that toward the end of a long achievement test in high school, boredom led him to finish out the test by creating random patterns on the bubble sheets. "It didn't affect anything, so why not?" Students who wish to attend college can almost always find some place that will accept them, irrespective of grade-point average.

Those going directly into the workplace know they need a diploma, but the quality of work is not a major factor; employers indicate that they seldom look at transcripts during the hiring process.

For all these reasons, advocates of the new accountability emphasize the need for "high-stakes" tests that *do* have consequences. When students do well, they should be promoted or graduated, and their teachers should earn a bonus. When students do poorly, they should not progress to the next level, and their teachers should lose their positions or their schools should be closed.

While the idea of consequences is intuitively appealing to the public, it remains the most controversial element in accountability, and most states have moved cautiously in offering either positive incentives or negative sanctions (Lynn Olson 1999c). Yet there is wide agreement that without sanctions, accountability will remain a hollow shell.

## Targeting Teacher Development

Newmann and colleagues note that accountability alone will not lead to improvement, but must be combined with organizational *capacity* to improve. Even in systems that provide strong incentives to succeed, teachers will not improve if they do not know how. Indeed, the effect of strong accountability in low-performing schools may simply be despair and depression. Thus, without the capacity for change, knowledge of poor results may undermine motivation rather than enhance it.

For that reason, accountability systems require a strong teacher-development component to build the needed capacity. Unfortunately, professional-development programs have often been an afterthought in many schools. Teachers get short-term exposure to a smorgasbord of topics, with little time and few resources available to follow through in depth.

An effective accountability system requires teacher-development activities that are tied to student achievement. Unsatisfactory test results should generate training and research opportunities aimed at bringing students closer to the goal. Ideally, these opportunities should be integrated into the daily life of the classroom, not just tacked on to the beginning of the school year.

## Unanswered Questions

Entering the new millennium, America's schools increasingly find that accountability is defining the nation's education agenda. The very first action of the George W. Bush Administration was to lay out an education proposal that incorporated many of the principles of standards-based accountability. Yet beneath the relentless forward movement, one could hear quiet voices raising questions that have yet to be answered.

### Accountability from Whom?

Advocates of greater accountability often evade the question of exactly *who* is accountable. Initially, the push for accountability pointed the finger at teachers who were seen as too self-satisfied, unfocused, or incompetent to get results. More recently, critics have conceded that the problem is more systemic than individual, and that teachers are often powerless to act (Gerstner).

But even if we accept that accountability is shared, we are left with perplexing questions about the responsibility of each participant.

*Teachers* generally concede their key role in the learning process, but balk at the idea that they can single-handedly produce student achievement. For one thing, learning is a complex, often mysterious process that goes on mainly in the head of the learner. To an outside observer, the link between cause (teaching) and effect (learning) is frequently tenuous and often invisible. While teachers often talk wistfully of seeing the light bulb go on over the student's head, they cannot always be sure what flicked the switch, nor can they count on being able to reproduce the effect at will. Education still lacks a universally accepted instructional paradigm. To teach a concept, we can rely on direct instruction, discovery learning, video presentations, kinesthetic activities, assigned reading, and many other possibilities. Each one works—some of the time, but not all the time.

More important, learning requires the cooperation and active participation of the learner, something not always forthcoming from students who are involuntary participants or who increasingly come to school with psychological baggage that makes academic learning a lesser priority. Jennifer O'Day (1997) notes that learning is a joint production of teachers and students. Teachers can plan, instruct, cajole, admonish, and evaluate, but in the end nothing happens unless the student accepts the challenge of learning. Likewise, Philip Schlechty (1990) sees learning ("knowledge work") as something that must be done by the student. The teacher's role is to establish conditions that enable and support the process, but the student must do the work.

Thus, teachers are more comfortable with an input model of accountability that defines their responsibility in terms of dedicated effort and use of "best practices" rather than outcomes that they don't directly control. In the minds of teachers, there is a lingering scent of unfairness about the idea of being held accountable for something beyond their immediate control.

*School leaders* are obvious targets for accountability, since virtually everything that happens in their buildings is considered in their domain. Most principals, in fact, are already familiar with—and resigned to—the idea they will be held responsible for outcomes that are beyond their direct control.

Despite this inherent accountability, many administrators, like teachers, are reluctant to frame their responsibility in terms of ensuring student outcomes. For example, the Association of Washington School Principals (http://www.awsp.org) endorses the idea that "the principal is accountable for the continuous growth of students and increased building performance," but then outlines specific responsibilities in terms of functions or inputs:

- Design, implement, and monitor building procedures and practices that promote a safe and orderly school environment.
- Advocate, influence, and sustain a school culture conducive to continuous improvement for students and staff.
- Lead the development, implementation, and evaluation of data-driven plan(s) for improvement of student achievement.
- Assist instructional staff in implementation of curriculum, instruction, and assessment aligned with state and local learning goals.

- Monitor and evaluate staff implementation of school improvement plans and effective instructional practice(s).
- Manage human and financial resources to accomplish student achievement goals.
- Communicate and partner with colleagues, parents, and community members to promote student learning.

In addition, AWSP points out the kind of support that must be provided by the district, as well as the authority principals require to carry out their responsibilities. Thus, as with teachers, principals are cautious about accepting unilateral responsibility for student learning.

*Students* clearly bear some responsibility for their own learning, a fact recognized by the states that have tied graduation to assessment scores. Yet how far can their responsibility be pushed? Students, after all, are minors, lacking the experience that allows them to make wise choices, and immature by definition. Can we ask them to achieve at a high level when they have not been offered the proper support, both at school and at home? This is especially true for students from disadvantaged backgrounds, for whom rigorous high-stakes assessments may be a one-way ticket to the margins of society.

*Parents,* by general consensus, play a key role in student learning, both by providing a safe, nurturing home environment and by supporting and reinforcing the work of the school. Unfortunately, teachers have seen too many students from homes where there is little willingness to provide for children's basic needs, much less a push to learn. Yet the accountability here is a moral rather than a legal one. Parents are required to send their children to school, and keep them from harm the rest of the time, but not much more than that. Indeed, market-oriented approaches to accountability view parents as customers whose only responsibility is to choose a school for their children.

Some schools have tried to develop parental accountability by having them sign "contracts" in which all participants have pledged to live up to certain responsibilities. More daringly, some schools are experimenting with "grading" parents by sending home checklists documenting how well students are prepared for school (Michelle Galley 2000). Yet these are mostly symbolic gestures, and it is not clear how schools could enforce any obligations on parents.

From all this, it takes no great imagination to make the case that accountability is collective, and that all stakeholders have a responsibility. Sarah Brooks (2000) says the key is reciprocity: "Feelings of accountability and responsibility arise from a mutual sort of agreement—you provide me with the tools and environment conducive to high performance, and I agree to do what I need to do to meet your expectations."

Yet in saying learning is *everyone's* responsibility, do we run the risk of undermining the sense of personal obligation that is at the heart of accountability? When we emphasize reciprocity, do we simply authorize finger-pointing? When we say, "students must do the learning," do we make it too easy for teachers to say, "I taught; they must not have learned"? Such questions have to be worked out through a continual process of dialogue and reflection, not through policy mandates.

### Accountability to Whom?

Wagner notes that in an accountability relationship, there is always someone to whom we owe our accounting. To whom are educators accountable? The most obvious answer is to whomever pays their salaries. Any job involves an implicit contract governing the exchange: a job to be done and a salary in return. Thus, governing boards and legislatures have the most obvious right to demand accountability.

However, most teachers and many administrators also see themselves as having a strong—perhaps even dominant—responsibility to their students. The abstract employment relationship is easily overshadowed by the immediacy and vibrancy of a roomful of needy students. Should teachers perceive a conflict between what they owe employers and what they owe their students, the resolution would be far from certain. For example, teachers may see a heavy emphasis on state standards and assessments as an injustice to students whose backgrounds have not prepared them to compete with more affluent students, or they may believe the assessments narrow the curriculum and drain it of richness and diversity.

In reality, most educators live in a web of responsibilities that make them accountable to many people for many things, and satisfying all those obligations requires continuous negotiation and dialogue.

### Accountability How?

Granted that educators owe an accounting to the designated authorities, what is the best way to ensure that they live up to their responsibilities? Much of the debate centers around motivation: What is the best way to keep students and teachers on track and fully living up to their responsibilities? The simple answer, which is embedded in so much of today's accountability debate, is that people respond to carrots and sticks. Reward good performance and punish poor performance, and they'll fall in line.

But while this view is intuitively appealing, psychologists can provide ample evidence that human motivation is infinitely complex, and sometimes counterintuitive. Teachers and students are not rats in a Skinner box, and they bring a world of experiences, perceptions, and values to the classroom. How they respond to the policymakers' rewards and punishments is not a foregone conclusion.

In fact, there are already signs of a "standards backlash," as a diverse group of critics take issue with key elements of standards-driven accountability (David Hoff 1999). Alfie Kohn (1993), for example, has argued that its motivational theory is too simplistic, and that standards often narrow the scope of the curriculum by exalting an outdated instructional strategy. Accountability, Kohn claims, "has approximately the same effect on learning that a noose has on breathing."

Other critics complain that the new accountability undermines professional autonomy, has a disproportionate effect on schools with large numbers of disadvantaged students, and attempts to hold teachers accountable for things they cannot control (Scott Willis 1999). More dramatically, students have protested the new emphasis on assessment by boycotting state-mandated tests (Jacques Steinberg 2000). Finally, no matter how compelling the rationale for standards-based accountability, it is still far too early to know whether it will have a significant effect.

**Accountability for How Long?**

Some educators harbor a suspicion that many people don't really *want* the high standards being demanded so glibly. It is easy to call for rigorous standards—Who could be against that?—but those standards may begin to chafe when failure brings the real consequences that accountability demands.

Occasionally the very policymakers who imposed a standards-driven system will back away from the ultimate implications. For example, after years of building a standards-based system, the Wisconsin legislature firmly rejected a requirement that graduation be tied to assessment results. Legislators raised questions ranging from fairness to expense, but may also have been envisioning the political fallout when schools began to deny diplomas to children of constituents. For their part, parents often lose enthusiasm for accountability if it derails their children's steady progress toward college or a decent job.

Such cases reinforce educators' fears that accountability will be one more short-term wonder, a fad that, like a tornado, sweeps down on a town and is quickly gone, leaving survivors to pick up the pieces.

All these questions have created a sense of caution in the education community. In January 2001, the Learning First Alliance, a broad-based coalition of professional organizations, called for "mid-course corrections" to standards-based accountability. In particular, the alliance articulated five core concerns:

*Alignment of standards, curriculum, and assessments.* Since teachers tend to teach what tests measure, assessments should be fully aligned with standards and should include higher level thinking skills using a variety of test formats. Likewise, schools should offer "deep and rich curriculum" that fully covers the spectrum of standards, not just language arts and mathematics.

*Adequate professional development for teachers and principals.* Successful implementation of standards requires practitioners to learn new skills, and "intensive and ongoing" professional development is essential.

*Sufficient resources and support for each child to meet high standards.* For all students to achieve high standards, states and districts must be prepared to invest significant resources in upgraded curriculum, improved training, extended learning time, smaller class sizes, modern facilities, and enhanced technology.

*Communication about the importance of standards and accountability.* States, districts, and schools must inform parents and public about the purpose, nature, and implications of standards-based accountability.

*Balanced and comprehensive accountability systems.* Student success should not hinge on a single measure of achievement. High-stakes decisions should be based on a broad range of indicators.

The alliance emphasized that it was not calling for a change of direction and expressed a "sense of urgency" about accelerating improvement efforts. Yet its statement, coming from unions, administrators, parents, teacher educators, school boards, and state education officials, signaled that the education community sees the route to accountability not as a straight-ahead charge, but as a cautious exploration of unknown territory.

## Managing the Accountability Challenge

Many school leaders have welcomed the current push for accountability because it underlines the responsibility of *all* stakeholders to focus on student learning. A unified push for achievement, with all parts of the system working in sync, is an energizing prospect. Yet the same leaders may also suspect that when the cheering is over, the buck will still screech to a halt where it always has: in the principal's office.

For now, school leaders are faced with the formidable challenge of integrating the new external demands into the life of the school, in a way that does not undermine the positive initiatives already under way at the local level.

The task brings both benefits and risks. On the one hand, external accountability provides a potent rationale for moving people off dead center. Most teachers understand the reality of state control and bureaucratic accountability, and recognize the need to respond, however grudgingly, to mandates. On the other hand, imposed standards can threaten local initiatives. Teachers who have worked hard to establish a rich array of authentic assessment tools in their school may be demoralized by having to measure student success with a state-mandated, multiple-choice exam. Seeing strong external controls, teachers may infer a lack of trust and a corresponding devaluation of their work (Willis). Helping schools find the balance between external and internal standards may be the principal's most critical task in the decade to come.

The remainder of this book explores the nature of that challenge. Chapter 2 explores the psychological assumptions behind the new accountability systems, drawing on motivational theory to identify the key factors that lead teachers, students, and parents to make achievement a priority.

The next five chapters examine the major components of the current accountability model and their implications for school leaders. Chapter 3 describes how state and local standards can create a publicly communicated, clearly understood set of expectations for learning. Chapter 4 discusses the appropriate use of assessment in the accountability system.

Chapter 5 looks at the critical role of incentives and consequences in motivating teachers, students, and parents. Chapter 6 provides recommendations for developing a system for reporting results to the public, and chapter 7 explains how professional development can support and enhance the accountability process.

# 2

# Understanding Human Motivation

*Can we be sure that a standards-driven accountability system will stimulate student and teacher effort in a productive way?*

Fresh out of college, the eager young employee repeatedly begged his boss for an assignment that would let him show his stuff. Finally, the executive called him in to announce that he was now in charge of a special project. The employee listened carefully, took notes, and asked a few questions. Then, expectantly, he asked, "If I do really excellent work on this project, what do I get?" The executive, a grizzled veteran of the corporate wars, gazed silently out the window for a moment, then fixed his protégé with a level gaze and said, "To keep your job."

Scan through any book on leadership, and you'll be almost certain to find a discussion of "motivation." From ancient Chinese sages to the quantum-science musings of Margaret Wheatley, analysts assume that leadership requires the ability to energize and unite others in the pursuit of worthwhile goals. Indeed, *leadership* is often defined in terms of the ability to motivate others, as in "a leader is someone who has followers."

So no one should be surprised when critics explain schools' perceived shortcomings as a failure of motivation. If students are not living up to their potential and teachers are not focused on student learning, it is because the system has failed to provide the proper incentives. Give participants a *reason* to achieve, and they *will* achieve. The current accountability movement is consciously designed to provide both the carrots and the sticks that will move the system off dead center.

For most people, this analysis is a matter of simple common sense, fully consistent with their knowledge of human nature, but it merits careful examination. If students and teachers are failing to achieve what society expects, there could be many causes besides lack of motivation.

John Keller (1999) notes that motivation is only one of three major influences on human performance. In addition to being willing to put forth the effort, people must have the knowledge and skills to do the job and must also be supported by an environment that offers the necessary resources. Perhaps teachers want to boost student achievement but don't know how; perhaps students want academic success but are trapped in schools that don't provide qualified teachers, adequate supplies, or a safe environment.

And even if we believe that the problem is motivational in nature, can we be sure that a standards-driven accountability system will stimulate student and teacher effort in a productive way? Loose talk about "incentives" may lead us to overlook the complexity of today's school environments, and school leaders gearing up to meet the accountability challenge may be wise to begin by asking what motivates students and teachers to behave as they do.

Motivation is an expansive topic, worthy of an entire book, so the discussion requires some boundaries. First, I will not attempt a comprehensive review of motivational theory but will attempt to extract from the literature some ideas and perspectives that seem especially relevant to accountability issues.

Second, I will deal in generalizations rather than try to provide a diagnostic instrument for understanding particular individuals. Although humans share many basic needs, motivational structures are complex and highly individual. A strategy that works for one teacher or student may be completely ineffectual with another.

Finally, I will not attempt to explain why some people are "motivated" and others are "not motivated." For one thing, there is no such thing as an unmotivated person; everyone, at every moment, is oriented toward some goal, even if the goal seems trivial or socially unproductive. When we describe people as unmotivated, we really just mean that they are disinclined to pursue goals that *we* believe to be valuable.

Instead, I will make an initial assumption that teachers and students are interested in being productive, happy, and socially responsible. Admittedly, contrary examples are not hard to find; for a variety of reasons, some people display motivational patterns that seem unproductive or harmful. But one has only to watch a first-grade class for a couple hours to sense the love of learning that children bring with them to school. And most teachers, most of the time, easily live up to Craig Pinder's (1998) definition of *work motivation*: the ability to get out of bed on a rainy Monday morning. Beyond that, many will work long hours, carry work home, and even spend their own money on classroom supplies.

So the real question here is not why some students and teachers are habitually apathetic about their work, but why all the human effort that goes into a typical school day so often fails to produce the results we want. How can school leaders build motivational systems that support and enhance the higher standards that are expected of today's schools? An examination of motivational theory may at least provide some clues.

In the remainder of this chapter, I will explore several perspectives on motivation drawn from the extensive literature on the subject, and will apply those ideas to the case of students and teachers. The first section will set the stage by discussing a fundamental theme in motivational analysis: the distinction between intrinsic and extrinsic motivation. The next section will explore the implications of three major theoretical perspectives on motivation, and the chapter will conclude by considering some of the special circumstances of teacher motivation.

## Motivating from the Inside, Motivating from the Outside

Psychologists have never come close to agreeing on a unified theory of motivation; explanations of human behavior are numerous, diverse, and frequently incompatible. But one common theme pervades the motivational literature: the debate over intrinsic motivation versus extrinsic motivation.

**Competence and Challenge**

Intrinsic motivation is a state in which people engage in an activity "for its own sake, for the enjoyment it provides, the learning it permits, or the feelings of accomplishment it evokes" (Mark Lepper 1988). People who are intrinsically motivated will enter an activity with no other reward required.

Edward Deci (1995), after decades of study, says that intrinsic motivation is based on a desire for competence and autonomy. He believes that people have a strong drive to gain mastery of their environment and to make their own choices, and prefer behaviors and activities they believe will lead to that result. Craig Pinder recalls a classic schoolboy recitation that expresses this idea:

> It matters not how strait the gate
> How charged with punishments the scroll
> I am the master of my fate
> I am the captain of my soul.

Although intrinsic motivation is thus tied to fundamental human needs, it can surface in activities that seem mundane or trivial: gardening, collecting stamps, backpacking, doing the Sunday crossword puzzle, trying to entice purple martins to nest in the back yard, and many other highly idiosyncratic pursuits. But no matter how mundane the activity, intrinsic motivation can be accompanied by moments of profound happiness. Deci, quoting Robert Henri, says it can lead to "a more than ordinary moment of existence." Abraham Maslow (1968) called such a moment a "peak experience"; Mihaly Csikszentmihalyi (1990) describes it as *flow:*

> We have all experienced times when, instead of being buffeted by anonymous forces, we do feel in control of our actions, masters of our own fate. On the rare occasions that it happens, we feel a sense of exhilaration, a deep sense of enjoyment that is long cherished and that becomes a landmark in memory for what life should be like.

Csikszentmihalyi notes that such moments are not simply a matter of "pleasure." The deepest sense of satisfaction often arises from the ability to transcend pain and frustration and to master a particularly difficult challenge.

What creates experiences like this? Csikszentmihalyi says that flow occurs when there is a task that is challenging, that calls up our best skills and energies, yet is doable. In other words, the task is matched to our current level of capabilities. He explains it as a balance between conflicting forces. If a task is too difficult, we become anxious and frustrated; if it is too easy, we become bored.

**Goals and Imagination**

In addition, intrinsically enjoyable activities generally have clear goals and provide unambiguous feedback. Knowing where we are headed and how much progress we are making frees us to concentrate on the activity. People involved in intrinsically motivating activities often report "losing track of time" or "forgetting everything else" because their attention has narrowed to the activity itself.

Interesting confirmation of this comes from Thomas Malone's (1981) study of video games. As any parent can attest, children find these seemingly mindless games motivating to the point of addiction. Why? Malone found four reasons:

1. *The goals are clear.* Success in video games requires highly concrete goals such as blowing up enemy spaceships or avoiding assorted hazards to rescue the princess.

2. *Feedback is instantaneous.* In a video game, you know *immediately* how you are doing (having your spaceship vaporized is an unmistakable sign that you did something wrong). Such quick feedback compels immediate corrective action, and the learning curve is accelerated.

3. *The level of difficulty changes to match the skills of the player.* The typical game is multistaged, consisting of a series of levels of increasing difficulty. The first level comes with training wheels: the puzzles are simple and easily solved, the enemy slow and dumb. When the player masters that limited challenge, the next level increases the challenge by providing

tougher puzzles and more formidable opponents. In this way, the game presents players with a series of ability-graded challenges that keep them fully engaged. (Not all games are equally motivational. Some are viewed with disdain, either because they are too simple—and hence boring—or because they jump too quickly to a high level of difficulty.)

4. *The game engages the imagination with an element of fantasy*. Players are not just chasing blobs of light around the screen; there is almost always a story line that has them participating in a quest or on a mission to save the universe.

Goals, feedback, optimum challenge, and imagination: taken together, those conditions allow complete concentration on the task at hand, untroubled by the usual mental distractions, and provide a brief "vacation" from everyday cares. When this occurs, an activity is seen as intrinsically satisfying, worth doing just for the sheer enjoyment of doing it.

### Ambivalence about Rewards

By contrast, extrinsic motivation is driven by the prospect of a reward or outcome that bears no direct relationship to the behavior. Whereas people will voluntarily engage in activities that are challenging and meaningful, they normally will *not* spend time on things that are boring, overly difficult, or distasteful, unless those activities can be "traded" for outcomes that *are* rewarding. In other words, while intrinsic motivation is an expression of personal desire or values, extrinsic motivation is a kind of economic transaction.

Not surprisingly, when people are extrinsically motivated, they view the activity in an impersonal, businesslike way, ready to disengage whenever the exchange fails to hold enough value. (This is why employers often have to pay premium rates to entice employees to work on holidays or weekends.)

Americans are profoundly ambivalent about extrinsic and intrinsic motivation. On the one hand, the idea of working for a reward is embedded in everyday psychology; no one is surprised that absenteeism declines on paydays or that a sales force can be energized by contests offering Caribbean vacations. So the idea that a diploma should be the reward for learning seems like common sense. Most teachers don't hesitate to distribute stickers for good behavior, and businesses proudly advertise incentive programs in which students can trade good grades for coupons. But at some point (the boundaries are not clear) reasonable "rewards" become inappropriate "bribes" in the eyes of many.

For example, the idea of paying students to attend school (as a few desperate school systems have done for at-risk students) appalls many onlookers, probably because schooling is a *gratis* service dedicated to the benefit of students; the idea of paying them to take advantage of it seems to violate a fundamental moral contract.

Moreover, extrinsic motivation, though accepted as a fact of life, is often held in lower esteem than intrinsic motivation, especially when it comes to school learning. Students, asked to recall a great teacher, usually pick one who "made learning fun." The adult public also seems to reserve its highest accolades for teachers who can make school meaningful for students. This can be seen in two recent movies about teaching.

In a highlight of *Dead Poets Society*, Robin Williams joyously asks his students to rip out the pedantic essay on poetry in their textbook, and tells his rapt class, in reverential tones, that poetry is about life and passion and that "the powerful play goes on and you may contribute a verse." In *Mr. Holland's Opus*, Richard Dreyfus makes a breakthrough with a struggling clarinet student when he tells her, "Music should be fun." Don't play the notes on the page, he advises; play the sunset. Teachers, it seems, are not celebrated for their well-designed token systems.

Arthur Powell (1996) notes that teachers themselves, influenced by a child-centered progressive tradition, use intrinsic motivation as the measure of their success. Powell cites the agenda:

> Identify individual student interests and build on them. Make curriculum relevant to students' lives. Make learning interesting, meaningful, and challenging. Give students more choices about what they should study. Emphasize creativity.

> Encourage them to work actively at learning by discovery, constructing ideas, asking questions, teaching peers. Make sure they understand the personal utility of whatever they are asked to do. Acknowledge their feelings when they find a class uninteresting. Give positive feedback in an uncontrolling manner. Allow everyone to feel success. Make education learner centered. Focus not only on understanding but on deep understanding. Abolish grades and all extrinsic incentives.

But this agenda, as appealing as it may be, holds risks. An uncritical embrace of intrinsic satisfactions may set teachers up for disappointment because it conflicts with some fundamental conditions of American education. Powell notes that the nation's wholesale commitment to compulsory mass education, combined with relatively weak incentives for achievement, does little to generate student enthusiasm for school. "Most students do not come to school mainly to learn. Their presence owes more to law, habit, peer relations, and the absence of anything better to do than the active seeking out of a valued service." He suggests that a judicious combination of intrinsic and extrinsic motivation, including high-stakes testing, may be the most realistic approach.

## Teacher Attitudes

When discussing their own attitudes toward work, teachers tend to speak the language of intrinsic motivation. Joseph and Jo Roberts Blase (1994) interviewed teachers who worked with successful principals and found that their satisfaction was expressed in terms of autonomy, empowerment, recognition, and appreciation. A typical comment by a teacher about the school's principal:

> She allowed the teachers to choose committees to participate in and to run those committees. Because I am able to have a say in how I think the schools should function, I feel empowered within our school. I like being trusted as a professional to make decisions concerning our school. My behavior could be described as confident. I feel more professional because I am being treated more professionally.

Blase and Blase suggest that appealing to intrinsic motivation may be the preferred strategy simply because school leaders have so little opportunity to offer extrinsic rewards (budget shortfalls and union contracts often preclude any thought of offering monetary rewards for good performance). Moreover, because of their attraction to intrinsic motivation for students, teachers may be more open to intrinsic rewards in their own teaching. For example, the authors noted that praise can be effective with teachers if it is sincere and not a transparent ploy to influence behavior.

For policymakers and school leaders, the words of Susan Mohrman and Edward Lawler (1997) probably represent the most accurate perspective: "The behavior of most people in employment situations is motivated by a combination of intrinsic and extrinsic factors." Given the magnitude of change required by standards-driven accountability, leaders cannot afford to overlook any possibility.

## Three Views of Motivation

We can easily imagine that somewhere on the African savannah, hundreds of thousands of years ago, an early human sat in the safety of a tree watching a companion face down a leopard over a scrap of scavenged meat and wondering, "Why does he *do* that?" From the beginning, understanding the actions of others—and sometimes our own—has been one of the central puzzles of the human condition.

In the last century psychologists have added sophistication and objectivity to a search for the answers, but as yet no clear consensus has emerged. Eons after that human ancestor confronted the leopard, psychologists can still engage in spirited debate about whether his behavior was driven by hunger, a desire for prestige, or the sheer intrinsic joy of outwitting a predator (the savannah equivalent of a high-stakes video game).

In this section I explore three viewpoints developed by psychologists in their century-long quest to explain human behavior. These perspectives are not tightly knit formal theories, nor are they mutually exclusive. Rather, each offers a kind of lens for looking at motivation, and each highlights dimensions of motivation that the others ignore. Taken together, they offer a useful tool for examining the assumptions of the accountability movement.

**Motivation as Satisfaction of Needs**

One of the earliest and most easily understandable views of motivation says that people act to fulfill certain basic needs. When hungry, they seek food; when cold, they seek warmth; when lonely, they seek companionship. People are willing to engage in any activity that satisfies such a need.

While "need" is actually a hypothetical construct in that it cannot be directly measured (Pinder), everyone has experienced it. The most obvious needs are physiological—food, water, oxygen, comfort. People who lack any of these are strongly motivated (if not frantic) to satisfy them. Beyond these self-evident requirements, psychologists have identified the following needs as major factors in human motivation.

1. *Relatedness*. Simply put, people need people, from simple companionship to lifelong partnerships. Social affiliation brings a wealth of satisfactions, including affection, attention, and support, and ranges in intensity from the desire to chat over an afterhours beer to the kind of love that leads people to sacrifice their own life to save another's.

Most organizations are built around elaborate social networks, and employees will interact with one another for purposes that go far beyond accomplishing the tasks at hand; the organization's agenda can easily take second place to employees' affiliation needs. This is equally true of classrooms, where the daily lesson plan is often subverted by students' needs to establish and maintain social relationships.

2. *Esteem*. Abraham Maslow, in his groundbreaking work on motivation, said that people have a strong need for "a stable, firmly-based, usually high evaluation of themselves, for self respect or self esteem, and for the esteem of others." Edwin Locke and colleagues put it even more strongly: "It is impossible for a human being to tolerate the full, conscious conviction that he is fundamentally no good, that is, evil,

worthless, inefficacious, without going insane or committing suicide."

Maslow stressed that the healthiest forms of esteem are built on *deserved* respect that comes from real competence, not just the opinions of others. As many teachers have found, soothing reassurances ("I know you can do it") are not as meaningful as the triumph of actual accomplishment.

Schools are designed to develop competence, providing students with multiple opportunities to enhance their esteem through achievement. But every opportunity to succeed is also an opportunity to fail, presenting students with a risky agenda. Esteem is always in play, and because it is such a strong need, students will maintain it in whatever way they can. Confronting the risk, some students opt out of the task (Deborah Stipek 1998); failing to try carries some negative consequences, but it at least allows the continued belief that "I could do it if I wanted." And even when students have conscious doubts about their own competence, they may work to disguise that doubt by creating the *illusion* of performance rather than working for actual mastery (Linda Lumsden 1999).

Teachers sometimes make similar choices. In a profession in which craft knowledge is filled with "endemic uncertainties" (Dan Lortie 1975), they may settle for well-worn worksheets rather than a risky open-ended discussion, or may preserve their self-esteem by saying there is not much that can be done when so many parents fail to support the learning process.

3. *Competence*. Anyone who watches a toddler try repeatedly to tie her shoes, climb a ladder, or operate the channel selector will be struck by the child's concentration, persistence, and exhilaration when success is attained. Mastering their environment is a major priority not only for children but adults as well. The strength and persistence of this urge, and the satisfaction it brings, has led some psychologists to list competence as an innate human need (Stipek, Lumsden).

However, not just any task will satisfy the competence motive; it must have the right amount of difficulty. "Optimal challenge is a key concept here," says Edward Deci. "Being able to do something that is trivially easy does not lead to perceived competence, for the feeling of being effective occurs spontaneously only when one has worked toward accomplishment."

Today's reform environment has engendered a set of standards that clearly meets the requirement of not being trivially easy, but it may challenge "optimum" from the other direction. Teachers who have honed their classroom strategies to the point of routine are now being asked to teach in unaccustomed ways; students are being asked to reach substantially higher levels of achievement. While the drive for competence is at least partly innate, it is also sensitive to the social environment, meaning that organizational policies and classroom practices can undermine it. Inadequate feedback, insufficient encouragement, and excessive competition can easily blunt the desire to gain mastery (Stipek).

4. *Autonomy*. We live in a world of limits; the things that we want most are not always available, and there are always forces pushing us in directions we prefer not to go. Understandably, the ability to control our own fate is highly valued, and some psychologists consider autonomy (self-determination) to be a fundamental human need.

Deci sees autonomy as operating in tandem with competence to produce intrinsic motivation. Simply setting out a challenge, without allowing the necessary freedom to develop the solution, will result in subpar performances. "Autonomy fuels growth and health because it allows people to experience themselves as themselves, as the initiators of their own actions.... A competent puppet does not nourish humanness."

In practical terms, a person's sense of autonomy is supported by being given a choice, which not only decreases alienation and pulls the person into the activity, but often leads to more workable solutions (Deci). The need for autonomy poses a major dilemma for standards-driven accountability, since the desired goals are typically chosen at the state level. Some policymakers have promised that the new constraints will be balanced by increased autonomy to devise strategies, but it is not yet clear whether this bargain is being kept.

5. *Curiosity*. As people explore the world around them, their attention is drawn to events that produce "surprise, incongru-

ity, complexity, or discrepancy from their expectations or beliefs" (Stipek). This kind of attention to novelty, which is observable even in very young infants, has obvious survival value, but also seems to express an inherent joy in learning.

Experienced teachers know that curiosity can be a powerful stimulus to learning. In *Dead Poets Society*, Robin Williams triggers this impulse at the beginning of his very first class by walking slowly to the back of the classroom and out the door, whistling softly to himself, and then sticking his head back in the room and saying, "Well, come on." Flabbergasted, his students follow him out into the hallway for a memorable lesson on life, mortality, and the importance of "seizing the day."

The exuberant curiosity exhibited by most preschoolers often seems to evaporate by fifth or sixth grade, a fact often attributed to the stifling effect of a too-narrow curriculum and methods. Curiosity does not proceed efficiently from Point A to Point B; instead, it meanders, finding its own path at its own pace. Faced with the pressure of meeting state-imposed standards through a multiple-choice test, both students and teachers may find curiosity to be something of a luxury.

6. *Self-actualization*. In his influential work on motivation, Maslow argued that when all other overt needs are met, people still engage in purposeful activity. He labeled this activity "self-actualization," and considered it to be a need in its own right. Unlike other needs, which temporarily subside when satisfied, self-actualization has an almost infinite capacity. "What a man *can* be," he said, "he *must* be."

In concrete terms, self-actualization was not a specific need but a kind of restless longing that could be expressed in many ways. In fact, what Maslow meant by the term overlaps with needs such as competence, autonomy, and curiosity. He attached a great deal of importance to self-actualization, claiming that those motivated by the need for personal fulfillment were happier and healthier than most people.

Although Maslow's theory has been highly esteemed by educators, its immediate value as a motivational tool is unclear, except as a reminder that humans are restless, striving organisms with an active personal agenda. That agenda may not match the immediate needs of the organization, but it does tell us that no one is unmotivated and that the leadership challenge is to hitch the organization's needs to the aspirations of its employees and clientele.

Whatever scheme is used to classify universal needs, there is little doubt that they exist, that they drive much of human behavior, and that they provide intrinsic satisfaction when achieved. Today's teachers and administrators, observing the increasing numbers of children who arrive at school hungry, scared, or emotionally neglected, are well aware that unfulfilled basic needs can undercut academic learning. More positively, leaders can fuel reform by viewing the change process as an opportunity to align the needs of the school with the needs of the people who work and learn there.

## Motivation as Reinforcement

Behaviorists offer the simplest theory of motivation, with the clearest strategies for application. People act in certain ways because their behavior is followed by personally rewarding consequences (reinforcements); behavior can be changed by arranging the environment to ensure that a desired behavior is followed by a reinforcement.

Unlike other motivational theorists, behaviorists have little interest in the inner workings of the mind. Leading advocates have always disdained efforts to explain behavior with the language of beliefs, feelings, and attitudes. In their view, it is much more efficient to proceed directly to the bottom line and apply two simple rules:

1. If a certain behavior is undesirable, remove any rewarding consequences.
2. If a certain behavior is desirable, arrange the environment so that a reinforcement follows the behavior.

Thus, if a student is disrupting class with noisy, off-task behavior, the teacher should make sure that the student receives no payoff for that behavior. For example, if the rewarding consequence is attention from the teacher or classmates, the teacher can ignore the behavior or can temporarily remove the child from the situation ("time out"). Better yet, the teacher should watch for the first sign of on-task behavior and make sure that it is immediately followed by a reinforcement.

Behavioristic thinking pervades American education, especially in elementary and special-education classrooms. Teachers at those levels routinely offer praise for desirable behavior, and some go further, offering stickers or other concrete reinforcers for good work, or even constructing elaborate "token economies" in which students earn tangible rewards for a wide range of specified behaviors.

Reinforcement theory is widely accepted because it echoes the everyday psychological theories that most people use; even those who have never heard of behaviorism know better than to give in to a child's tantrum. Even more important, behavior modification gets results, and teachers reflexively embrace whatever practical strategies will get them through the day.

Nevertheless, behaviorism has always attracted critics, for at least three reasons: concerns over practicality, doubts about the educational impact, and disagreements over philosophy.

1. *Practicality*. Early behaviorists, such as B.F. Skinner (1972), built their theory around experiments with rats and pigeons, whose motivational structures are presumably simpler than that of the average eight-year-old. Modifying the behavior of individual lab rats in a cage turns out to be much easier than dealing with a couple dozen "rug rats" in a highly interactive classroom.

For example, behaviorist advice usually begins with an admonition to "chart the baseline behavior," but few beleaguered teachers have time for such a step. Similarly, a common behaviorist prescription for off-task behavior is to ignore it—a response that many teachers find risky.

More important, finding the right reinforcement is largely a matter of trial and error. While everyone responds to reinforcement, not everyone finds the same things reinforcing. For example, teacher praise becomes steadily less rewarding as children grow older. If a first-grade teacher says, "I really like the way Nicholas is sitting up so nice and straight," the entire class will shift position to emulate Nicholas. If a seventh-grade teacher says it, Nicholas will do his best to shrink *into* the seat.

Finally, administering reinforcements requires a certain precision that isn't always possible under classroom conditions. One teacher recalls trying to reduce off-task behavior by giving a student a few pieces of sweetened cereal every time he returned to his desk. After a few rounds of this, the student figured out that he could increase the number of times he was rewarded for returning to his desk by increasing the number of times he *left* his desk.

2. *Educational impact.* A growing number of critics argue that even when behavior modification works in the short run, its long-range effects undermine some of our most cherished educational goals. In particular, critics argue that reliance on extrinsic rewards robs learning of intrinsic satisfaction. Deci's numerous studies lead to the same conclusion: Using extrinsic reinforcement for a behavior reduces its intrinsic satisfaction. Even when his subjects had previously found intrinsic value in an activity, they lost interest after the extrinsic rewards came to an end. When people treat an activity as a way of making money, they lose the excitement they once found in it. Thus, teachers may gain short-term compliance with extrinsic rewards, but at the cost of reducing student interest in the material.

Moreover, when people view an activity only as a means to a reward, they may reshape it to enhance the payoff. As Alfie Kohn asks, when children are given pizzas for reading books, how many are likely to choose long, difficult books? Stipek cites a study by James Gabarino in which sixth-graders served as tutors for first-graders. The tutors who were paid for their work were more demanding and critical of their students (often creating a negative atmosphere), presumably because the reward narrowed their attention to performance issues.

3. *Philosophical disagreements.* B.F. Skinner, the father of modern behaviorism, cheerfully conceded that his approach was highly controlling. He saw that as a plus: If we believed a certain goal was valuable, then why not reach it in the most direct and efficient way possible? Critics have not been so willing to dismiss the issue, arguing that behaviorism is inherently manipulative and even undemocratic.

Kohn points out that rewards are based on an inherently asymmetrical relationship: The person offering the reward has the power. Even something as innocuous as praise implies a power differential; the praiser is presumed to have superior knowledge and authority. (Of course, in a teacher-student relationship the relationship *is* asymmetrical, but if the long-range goal is to encourage critical thinking and student independence, a steady diet of praise can be counterproductive.)

Kohn also argues that "rewards rupture relationships," especially if the rewards are scarce. A common argument against merit pay is that teachers would be more inclined to keep their best ideas to themselves rather than giving away a competitive advantage. At a time when collaboration is a key priority for many schools, extrinsic rewards may add a divisive element.

### Motivation as Cognitive Attribution

Hard-core behaviorists such as B.F. Skinner were adamant that behavior could be influenced without worrying about thoughts and feelings, which were regarded as irrelevant *effects* of behavior rather than *causes* of behavior. One need only apply the right reinforcement and the desired behavior would follow.

But outside the easily controlled world of the laboratory, even behaviorists spend considerable time analyzing the inner motives of those they are trying to change. When seeking to influence another person, we seem to instinctively put ourselves in the other's place, asking, "If I were him, what incentive would encourage me to do this?"

How people perceive the world indeed influences the way they behave, sometimes in ways that initially seem counterintuitive. For example, in the 1970s, psychologist Mattina Horner (1972) identified a phenomenon in some women that she called "fear of success." Horner found that when these women were shown images suggesting career success, they called up other images suggesting negative consequences, such as "loss of femininity." Thus a presumably attractive outcome turned out to be a source of conflict and ambivalence.

What Horner found may no longer be true in today's social climate, but her work serves as a reminder that people have a rich inner life that regularly influences their behavior. Psychologists have recognized this fact through a cluster of theories that generally go under the heading of "cognitive attribution theory" or "expectancy theory" (Stipek, Pinder). At the heart of these theories is this belief: The value of an incentive is not in its objective qualities but in what people think of it.

For example, many purported incentives are not wholly positive. Just as Horner's subjects saw negative possibilities in a seemingly desirable goal, many possible outcomes lead to ambivalence:

- the raise that puts one in a higher tax bracket
- the high grade that raises parental expectations
- the promotion that leads to longer hours and greater stress

Thus the motivational value of an incentive will reflect the balance of positive and negative values attached to an outcome.

Expectancy theory also tells us that an incentive will be motivational to the extent that we see a connection between performance and payoff (Pinder). That is, offering extra money for meritorious performance will have an effect only if employees *believe* that the money will be forthcoming. One only need read the popular "Dilbert" comic strip for a few weeks to recognize the depth of cynicism that pervades many workplaces. For many employees, it takes a real act of will to believe that rewards will be issued or that they will bear any relationship to performance. The same is true of many students, who see only a dim connection between classroom performance and the promised rewards of real-life success. In some cases, the payoff is too far in the future to be meaningful, and in other cases students simply lack examples of older students who have parleyed a good school record into career success.

In addition, intended incentives are motivational only if the individual believes that he or she can (or will) perform to the required level. A million-dollar reward for making the Olympic swimming team would utterly fail to stir the 99.9 percent of Americans who know their skills are below the required level.

This last idea is well entrenched in the way teachers think about student motivation, usually expressed in terms of "self-esteem" or, more accurately, "self-efficacy." (*Self-esteem* is a broad term expressing the degree to which individuals take pride in their status and achievements; *self-efficacy* refers to the confidence that students bring to a particular task.) While some critics believe that an undiscriminating emphasis on self-esteem leads to a dumbing down of academic achievement, research shows fairly clearly that higher degrees of self-efficacy have significant effects on learning, including higher performance, greater strategic flexibility, and more realistic self-evaluations—regardless of actual skill level (Albert Bandura 1997). In addition, students with high self-efficacy are likelier to persist when they encounter difficulties.

The same applies to teachers, who have differing beliefs about their ability to influence student performance. Teachers with high self-efficacy believe that even difficult students are reachable with sufficient effort and instructional flexibility; teachers with low self-efficacy believe there is little they can do to help unmotivated students with unsupportive parents.

These beliefs, in turn, affect the way teachers go about their business. Those with high self-efficacy devote more classroom time to academic activities, offer extra help to students, and praise student accomplishments. Those with low self-efficacy spend more time on nonacademic activities, more readily give up on students, and are more likely to criticize students (Sherri Gibson and Myron Dembo 1984). Summing up the research in this area, Bandura (1997) says that teachers with low self-efficacy tend to favor a custodial orientation to teaching.

Thus, predicting the effectiveness of an incentive requires the exercise of a complex motivational calculus involving the degree to which students or teachers value the incentive; the degree to which they believe that the prerequisite performance will lead to the desired outcome; and the degree to which they believe they can perform to the required level. Doubts in any of the three areas will undercut the power of the incentive (Pinder).

To illustrate the importance of cognitive attribution in motivation, consider how students might respond to a high-stakes test, such as one required for graduation. To be motivated to perform well on the test, students must make three mental calculations.

First, students need to find value in the incentive. For most students, test-taking is not intrinsically motivating, so the incentive must be the earning of a diploma. We can reasonably assume a diploma has some value for most students, if only because they view it as the passport to other benefits, such as a job with a livable wage. But for some students the family history or community context may devalue the importance of the diploma, either because few family members have achieved that goal or because the community offers decent jobs for those lacking a diploma. In addition, for some minority students, academic achievement may be derided as "acting white."

Second, students must see the connection between test performance and the diploma. Unless they are profoundly cynical about the educational system, most students are likely to accept the link. However, some students may have more difficulty seeing the link between the diploma and career success, especially if their daily walk to school takes them past unemployed high school graduates sitting on the street corner.

Finally, students must believe that they can achieve the necessary level of performance on the test. This may be the biggest barrier for many students. Because high-stakes tests are comprehensive, covering many subjects, students may feel it is pointless to try to prepare for the tests. Then there is the fear factor. Students who have scraped through the system by dint of effort, charm, or generous grading policies may be intimidated by the knowledge that their whole future is riding on three hours with a number 2 pencil. Whatever their actual skill level, anxiety is likely to diminish their performance.

Thus, what seems like a simple common-sense incentive does not automatically stoke the motivational fires. An effective incentive system requires acute insights into the thoughts, feelings, and values of those at whom it is aimed.

The three perspectives we have just examined offer a well-rounded look at human motivation. A consideration of basic human needs reminds us that people do not approach work with a blank slate, but always bring a personal agenda with them. Behaviorism encourages us to think in terms of payoffs, recognizing that people will seldom persist in behaviors that do not lead to a personally gratifying result. Cognitive attribution theory forces us to recognize that it is not just incentives that matter, but the *meaning* that people attach to those incentives.

Unfortunately, these three viewpoints do not point to a clear or straightforward path for reforming schools through accountability. If anything, the theories remind us that people are complicated, especially when their lives are woven into a larger social and institutional fabric. However, these ideas also suggest that there are many ways people *can* be motivated, and with careful attention to the needs of teachers and students, accountability can become an integral part of school life. The next section applies some of these insights to the work of teachers.

## Motivating Teachers

The first part of the chapter discussed "human motivation" as though all humans were the same. In reality, of course, people may share certain universal needs, but they differ considerably in how they seek to meet those needs. While an examination of all the possibilities is far beyond the scope of this book, it may be worthwhile to consider whether teachers, as a group, have some common motivational structures that may influence their response to accountability initiatives.

We can begin with a stereotype that appears to be accurate: Teachers are service-oriented. They enter the field knowing full well that high incomes and social prestige are not part of the reward structure, and, though they may regret that fact, they appear ready to find their satisfaction in service to students (Robert Serow 1994). Dan Lortie, in his classic study of the lives of teachers, found that teachers were motivated primarily by "psychic rewards." That is, what kept them going was the light bulb switching on over the student's head or students coming back three years later to say how much they had appreciated the class.

Few of the teachers Lortie interviewed seemed satisfied with having students achieve academic mastery; virtually all expressed goals for their teaching that went beyond mere mastery of the official curriculum. Some of the goals were moral:

> You have to prepare them for life. I don't care if they don't know how to typewrite, they have to be individuals first. They have to be respectful... honest and respectful... good citizens and so forth.

Some of the extra goals were focused on connecting students with the wider world of learning, and developing a love of learning:

> Instill a love of learning from within. Not learning for the sake of bettering one's economic status but the love of education for the sake of education. The enrichment of the individual's life.

A third category of goals was universalism: the desire to reach *all* students:

> I'm trying to get every kid to be able to read as well as he can. Until every kid that I touch can read what he's supposed to, I'm not happy.

Lortie also found that when teachers talked about "craft pride" (the occasions that gave them greatest satisfaction), they focused on events with a strong personal element. Many recalled the "spectacular case," in which the teacher's determined effort finally broke through to a difficult student, or they mentioned the affirmation that came from former students coming back and expressing their appreciation. It is interesting that only a minority of teachers cited general academic success—such as good test scores—as a source of craft pride. Lortie noted, "It is as if they are uncertain of the tangibility of measured gains or the rightfulness of their claiming credit for them."

A partial reason for teachers' emphasis on psychic rewards is that extrinsic rewards are in short supply. Lortie noted that teaching is "unstaged," meaning that the progression from novice to expert is not outwardly visible through raises or promotions. Teaching has a flat career trajectory in which a masterful teacher, after twenty years of experience, still has the title "teacher" and is making the same salary as a mediocre twenty-year veteran. The connection between effort and reward is tenuous, and teachers do not have the sense of a "dues-paying" period in which sacrifice and hard work will ultimately be followed by a reward.

Partly for that reason, teaching has a strongly egalitarian outlook that makes extrinsic rewards largely irrelevant (Patricia Wasley 1991). In addition, Lortie noted that teachers found it difficult to articulate craft knowledge; teaching was marked by "the absence of concrete models for emulation, unclear lines of influence, multiple and controversial criteria, ambiguity about assessment timing, and instability in the product."

When a teacher has a breakthrough moment with a student, it is often difficult to say why. What was done differently on this occasion than previously? Because the answer is so often elusive, many teachers are uneasy about publicly distinguishing good teaching from poor teaching or about being evaluated by someone who is not familiar with their classroom. Some teachers resolutely maintain that no one can decide who is a better teacher.

Because of this set of beliefs about teaching, teachers are adamant about equal treatment. Few unions are willing to consider any element of subjectivity in determining salaries, so in most districts compensation continues to be based solely on longevity and schooling. Any kind of special recognition for some teachers is viewed with distrust and skepticism. Some teachers actually shy away from external awards because of the negative response from colleagues (Ann Bradley 1995).

None of this means that extrinsic motivators such as money are futile. In Richard Brandt's (1990) study of teachers participating in career-ladder programs, many teachers said that the added money was the main reason for their participation, and in recent years a growing number of teachers have expressed interest in certification from the National Board of Professional Teaching Standards and the cash incentive that sometimes goes with it. So teacher behavior *can* be influenced by the right extrinsic reward.

The surest route to a teacher's heart, however, is through the intrinsic satisfactions he or she finds in the classroom. This might be done in a number of ways:

1. Frame reform efforts in terms of "reaching students," rather than the abstract need for improved test scores.

2. Streamline the policies and procedures that get between teachers and their students. Teachers who get their deepest satisfactions from helping students inevitably get frustrated by the dozens of daily distractions that keep them from pursuing that goal. Lortie notes, "Teachers want to teach." Whatever allows them to do so will be seen in positive terms. Ironically, teachers who decline to participate in career-ladder programs often do so out of fear that the added paperwork will divert their attention from their classrooms (Brandt).

3. Help teachers see the connection between state-imposed standards and their own visions of good teaching.

4. Where monetary incentives are available, use them to add resources at the school level rather than having teachers compete for personal compensation.

These suggestions are just a quick sketch of the kind of recommendations that result when teachers' motivational structures are thoughtfully analyzed. The remaining chapters will continue this analysis in more depth (for students as well as teachers) as we explore the key elements of accountability systems.

# 3

# Educational Standards

*The emphasis on standards reflects an emerging consensus on how to reform schools.*

From the beginning, standards have driven the school-reform movement. *A Nation at Risk,* the report that started it all in 1983, grabbed headlines with its metaphor of "a rising tide of mediocrity," an image that crystallized public apprehensions of declining academic rigor.

Initially, the call for standards was just a vague ideal, a way of saying schools should aspire to a state of excellence that was unspecified, but surely better than the current reality. As the reform movement gathered steam, however, standards were increasingly defined in terms of tangible, explicit outcomes.

Today, standards are omnipresent. Robert Marzano and John Kendall (1998), after a comprehensive survey of state and national initiatives, characterized schools as being "awash in a sea of standards." They reported finding 200 distinct standards accompanied by 3,093 specific benchmarks and estimated that adequately addressing all these standards would require students to attend school through grade 22.

The proliferation of standards has been accompanied by wide public support. Overwhelming majorities of parents (83 percent), teachers (79 percent), and employers (94 percent) say that guidelines for student learning will improve academic performance (Public Agenda 2000). While some observers see signs of a "standards backlash," for the moment the standards movement thoroughly dominates the educational landscape.

Clearly, standards feed on a widespread public dissatisfaction with school achievement. Over 88 percent of Americans express concern about "low academic standards" (Peter Hart survey cited by Public Agenda), while media pundits and politicians fill the airwaves with references to troubled schools, social promotion, and functional illiteracy.

Some of the push may reflect self-interest. Given the strong public support for standards, legislators have every reason to mandate comprehensive accountability systems. Businesses believe that low standards push up their operating costs because of the need for remedial training, extra screening, and high turnover (Nelson Smith 1996). And some educators may see standards as a way of ensuring attention and resources for their subject specialty. During hearings for the National Education Commission on Time and Learning (1994), a representative of the National Geographic Society observed, "Implementing our standards will require more time. Geography is hardly taught at all in American schools today." An arts-education advocate put it more strongly: "I am here to pound the table for 15 percent of school time devoted to arts instruction" (National Education Commission on Time and Learning).

But above all else, the emphasis on standards reflects an emerging consensus on how to reform schools. After decades of limited success with enhancing inputs such as money, time, and training, policymakers have put their faith in an outcome-driven system anchored by clear standards. Rather than using disjointed, scattershot initiatives in curriculum, assessment, teacher education, and professional development, reformers are betting that an uncompromising adherence to standards will improve student performance (American Federation of Teachers 1999).

The logic is straightforward: When goals are expressed as unambiguous standards, educators will begin to align curriculum and instruction to serve those goals. Clear standards allow educators and students to focus on learning without having to ask, "What's important? What comes first?" Much like a market economy in which human effort is organized by the drive for profit, a standards-driven school will unleash all kinds of effort and ingenuity in pursuit of excellence.

## Types of Standards

For all these reasons, standards are firmly embedded in the language and policies of today's education-reform movement. Despite the hopes of many advocates, however, standards have not led to standardization. Instead, the accountability movement has resulted in an explosion of competing standards, and the term itself means different things to different people (Don Burger 1995). A survey of the accountability literature finds references to at least eight kinds of standards.

### Goals 2000

For the last decade, federal efforts to promote standards have been centered on a broad set of national goals articulated at the first education summit in 1989. These goals (the initial six were later expanded to eight) envisioned that the following would be accomplished by the year 2000:

1. All children will start school ready to learn.
2. The high school graduation rate will increase to at least 90 percent.
3. All students will become competent in challenging subject matter.
4. Teachers will have the knowledge and skills that they need.
5. U.S. students will be first in the world in mathematics and science achievement.
6. Every adult American will be literate.
7. Schools will be safe, disciplined, and free of guns, drugs, and alcohol.
8. Schools will promote parental involvement and participation.

These goals have generally been treated as worthy aspirations rather than true standards. Few people expected them to be met by the target date, but they have served as the platform for federal efforts to promote standards, resulting in measurable progress on a number of fronts (National Educational Goals Panel 1999). For a while the goals led to lengthy debate over the prospect of common nationwide academic standards. Because of widespread opposition to federal mandates, policymakers have handled the issue gingerly, suggesting that any federal standards would be voluntary. More helpfully, the 1994 Goals 2000 act provided substantial funding for state and local efforts to develop standards.

### Content Standards

Content or subject-matter standards, sometimes called "outcomes" or "exit goals," describe what students should know and be able to do as a result of their learning experiences (Ivor Pritchard 1996). Content standards are the driving force in the current accountability movement and are widely regarded as the key to school reform. They proliferated in the 1990s at both the national and state levels.

Subject-matter standards identify what students should know and be able to do in different disciplinary areas. During the past decade, nearly every subject-area group with an interest in K-12 education has formulated a vision of what students should learn in that field by the end of high school.

Even though these are national standards, they have rarely become part of public policy or formal accountability systems. A few, such as the standards developed by the National Council of Teachers of Mathematics (2000), have been treated respectfully and have made significant inroads into classroom practice. Some, like the history standards proposed by the National Center for History in the Schools (1996), have been vigorously attacked because of their controversial content. Others, like the English standards offered by the National Council of Teachers of English (1996) and the International Reading Association, have been politely acknowledged and then ignored.

Subject-matter standards have undoubtedly had an indirect effect on the development of state standards, but thus far their implementation has depended on the professionalism and enthusiasm of teachers rather than on district or state mandates.

State content standards have been developed by state departments of education or blue-ribbon commissions to provide a framework for statewide education reform. Currently forty-nine states have such standards or are developing them. (Iowa, as a matter of state policy, leaves curricular decisions to local districts.)

Across the country, what is defined as *content* covers a wide spectrum of goals. Standards are usually organized into major subject areas, particularly language arts and English, math, science, and social studies. Within those divisions, the standards can be organized in diverse ways that are subjective and almost "idiosyncratic" (Marzano and Kendall 1996).

Douglas Harris and Judy Carr believe that standards fall into three categories. "Essential knowledge" focuses on key ideas and concepts; "skills" are strategies for thinking, working, communicating, and investigating; and "habits of mind" are broad capacities such as developing satisfying relationships and self-evaluation.

Marzano and Kendall (1997) also use three categories of content standards: "declarative," dealing with facts, concepts, vocabulary terms, time sequences, cause/effect sequences, events, generalizations, and principles; "procedural," comprising skills, strategies, and processes, including algorithms, strategies, and macroprocesses (complex skills such as reading and problem-solving); and "contextual," involving knowledge and skills attached to a particular context (for example, modeling numbers using a number line). However, not all standards fit neatly into one category or another.

Because content standards are usually stated as broad, long-term outcomes of a K-12 education, they can seem formidable to teachers. When the standard says, "students will apply basic processes of logical reasoning," what are the implications for a third-grade teacher? To answer such questions, standards are often broken down into *benchmarks*, which indicate the optimum progress at particular ages.

For example, one math standard in Washington State asks that students "use mathematical reasoning to draw conclusions and verify results." The benchmarks for that standard indicate that by fourth grade, students should be able to "support arguments and justify results based on own experiences"; by seventh grade the expectation has been raised to "support arguments and justify results using inductive reasoning"; and by tenth grade the goal is to "support arguments and justify results using inductive and deductive reasoning."

### Performance Standards

Content standards typically incorporate complex and sophisticated kinds of knowledge and skills. For example, a typical geography standard says, "understands how human actions modify the physical environment." Although the general intent is clear, how do we know when students have met this standard? How well must they know it?

To answer questions like those, most frameworks include some kind of performance standards, which describe the different degrees of achievement (Diane Ravitch 1995). For example, Marzano and Kendall (1996) point out that student performance on a writing task might be judged as "advanced" (demonstrates precision in word choice), "proficient" (demonstrates adequate word choice), "basic" (word choice is adequate but limited and occasionally vague), or "novice" (word choice is limited and immature).

A key decision when using performance standards is determining which level of performance will count as meeting the standard. More often than not, policymakers choose "proficient" as the targeted level. While there is considerable rhetoric about "world class" standards, Marzano and Kendall (1996) caution policymakers that "world class" may be overly ambitious. They suggest that academic learning may be like swimming, where 1,000 yards of daily practice produces 75 percent of maximum attainment, and 2,000 yards daily produces 85 percent of maximum, but getting to 95 percent of maximum takes 10,000 yards daily. In other words, the effort and resources needed for peak performance increase geometrically. In addition, "world class" currently has no clear definition (Council of Chief State School Officers 1997).

### Skill Standards

This term is most commonly used to describe workplace skills. In 1991, the U.S. Department of Labor issued a report outlining the skills and personal qualities needed to succeed in fast-changing high-performance workplaces (Secretary's Commission on Achieving Necessary Skills 1991). The report identified five broad competencies:

- Resources: identifies, organizes, plans, and allocates resources.
- Interpersonal: works with others on teams, teaches others, serves clients, exercises leadership, negotiates, and works with diversity.
- Information: acquires, organizes, interprets, evaluates, and communicates information.
- Systems: understands complex relationships and can distinguish trends, predict impacts, as well as monitor and correct performance.
- Technology: works with a variety of technologies and can choose appropriate tool for task.

A number of states and districts have used these guidelines to develop more detailed standards, sometimes related to specific occupations (Bettina Lankard 1995). These have been used mostly in vocational and tech-prep programs, but some agencies have also attempted to integrate these work-oriented standards with academic standards.

## Curriculum Standards

Although the term "curriculum standards" appears occasionally in the accountability literature, it does not have a clear or consistent meaning. Some authors use it as an apparent synonym for "content standards," but even a detailed set of standards does not compose a curriculum. Standards specify an outcome, an end point; curriculum is a vehicle for getting there.

In theory, for every set of standards, there is a wide range of curricular choices that could be made. Thus, if a standard calls for the ability to analyze and interpret literature, works by Mark Twain, Feodor Dostoevsky, or Jorge Luis Borges could equally well serve the purpose. However, if standards specify extensive lists of essential knowledge, this flexibility disappears and the standards begin to look like a curriculum. Standard-writing always creates a certain amount of tension between those who believe that mastery of large amounts of knowledge is the mark of an educated person and those who prefer to emphasize broad-based thinking skills. For example, the Thomas Fordham Foundation has criticized many state standards because of their failure to include specific material that the foundation believes to be essential (Finn and Petrilli 2000).

In general, there are no widely recognized standards for curriculum. Phi Delta Kappa has offered five criteria that it uses when performing "curriculum management audits" for school districts:

- control of resources, programs, and personnel
- establishment of clear and valid objectives for students
- internal consistency and rational equity in program development and implementation
- use of the results from district-designed or -adopted assessments to adjust, improve, or terminate ineffective practices or programs
- improved productivity

From an accountability perspective, the key relationship between standards and curriculum is *alignment*. If the prescribed curriculum offers a plausible pathway for helping students attain the desired standards, it is appropriately aligned to those standards.

## Opportunity-To-Learn Standards

As states ratcheted up their performance expectations for students, some critics pointed out that simply upgrading standards, without providing support, would be a sink-or-swim approach in which disadvantaged students were at greatest risk (Richard Elmore and Susan Fuhrman 1995). They argued that students should not be held accountable for performance unless schools are held accountable for providing the necessary support. In practice, such support has usually meant guarantees about adequate funding, teacher quality, class size, and other traditional input measures.

For that reason, the opportunity-to-learn concept is often seen as running against the grain of the current results-oriented accountability movement. The two concepts do not necessarily contradict each other, however. Andrew Porter (1995) points out that opportunity-to-learn can reinforce outcome-based accountability if it focuses on quality of instruction rather than on quantity of resources. For example, many new standards ask students to acquire critical-thinking skills not attainable through traditional instructional methods; an opportunity-to-learn standard that identified appropriate methods for teaching the new skills could be a useful part of an overall accountability system.

### Teaching Standards

Teaching standards specify the instruction, activities, and projects teachers provide to help students attain content standards. Because standards-driven accountability focuses on outputs rather than inputs, teaching standards have played a secondary role in recent years. The logic of the system is that policymakers set the end results, which teachers attain by using whatever methods work best.

However, teaching can also be viewed as an element in opportunity-to-learn standards. When new expectations ask students to think critically and form deep conceptual understandings, schools may be forced to rethink their instructional methods. For example, Rhode Island has identified eleven propositions about teaching that specify the kinds of teacher behaviors that lead to learning. One states that "teachers create instructional opportunities to encourage students' development of critical thinking, problem solving, and performance skills." This proposition is further broken down into strategies such as the following:

- Design lessons that extend beyond factual recall and challenge students to develop higher level cognitive skills.
- Pose questions that encourage students to view, analyze, and interpret ideas from multiple perspectives.

- Make instructional decisions about when to provide information, when to clarify, when to pose a question, and when to let a student struggle to try to solve a problem.
- Engage students in generating knowledge, testing hypotheses, and exploring methods of inquiry and standards of evidence.
- Use tasks that engage students in exploration, discovery, and hands-on activities.

## Teacher-Education Standards

As a number of thoughtful observers have pointed out, the new standards require teachers capable of helping students achieve at higher levels than ever before (Linda Darling-Hammond). Because of this, and because teacher education has been persistently held in low esteem, many states have moved toward upgrading teacher-education requirements.

Some of the reforms have taken the traditional path of increasing admission requirements or adding coursework, but an increased emphasis on demonstrated performance is also apparent. The National Council for Accreditation of Teacher Education is implementing a performance-based system linked to assessments developed by Educational Testing Service (Ann Bradley 1999).

Simultaneously, the National Board of Professional Teaching Standards (1989) has developed advanced certification centered on five propositions:

- Teachers are committed to students and their learning.
- Teachers know the subjects they teach and how to teach those subjects to students.
- Teachers are responsible for managing and monitoring student learning.
- Teachers think systematically about their practice and learn from experience.
- Teachers are members of learning communities.

Although these propositions are stated rather abstractly, the board has translated them into very specific and rigorous tasks that candidates must accomplish at a highly proficient level.

It is apparent to proponents and critics alike that the many varieties of standards do not yet constitute a tightly knit, coherent system. Reformers, nonetheless, are obviously converging on the idea that clear and rigorous standards will generate lasting change.

The remainder of this chapter explores the implications of content standards for K-12 schools. The next section contrasts the logic of standards with the culture of teaching, and the final section examines how standards are chosen, the instructional changes the new standards will require, and the support systems that must be in place for standards to have an impact.

## Standards and the Culture of Teaching

For policymakers, implementation of standards is a straightforward process: identify critical knowledge and skills, develop appropriate assessments, and provide incentives. For educators, standards provide more questions than answers. Kate Jamentz (1998) notes that the usual rhetoric of standards does not respond to the serious questions that most practitioners ask. Whose standards should these be? How do we develop the capacity to carry them out? Do we really mean "all students?" What exactly is a standard? And what are the "right" ones? What will this have to do with my work?

Jamentz argues that a mere listing of standards, no matter how sincerely generated, will not produce significant change. While schools often claim they "have" standards when they can point to a document, the standards are most meaningful when viewed as a "call to action," a statement of what is worth fighting for. (Diane Ravitch notes that one older meaning of "standard" is a pennant or banner that goes out front and serves as a rallying point.) From that perspective, standards do not represent fixed targets but instead serve as a source of dialogue about what's important.

Students in a true standards-based system must behave very differently than in the traditional system. They must be actively engaged in meaningful work, must be able to describe what is expected of them and how they can achieve it, and must know when to ask for assistance. For students to do that, teachers must also change, designing activities aligned with the standards, analyzing (not just scoring) student work, making fair judgments of quality, and teaching students how to evaluate their own work. In turn, schools must also change, articulating a clear purpose, supporting norms of dialogue and collaboration, developing a commitment to internal accountability, and allocating resources appropriately. These are all deep changes, requiring "profound shifts" from traditional practice and requiring a "new mental model" (Jamentz).

On the surface, teachers appear to be giving qualified support to standards. In one survey, 87 percent of public-school teachers said that raising academic standards was a move in the right direction, though only 32 percent said it was "very much" in the right direction (Belden Russonello and Stewart).

Almost three-quarters of the teachers surveyed felt that the current level of standards in their state was set at an appropriate level, whereas 17 percent believed the standards were set too high. A majority of teachers reported that recent years had seen a more demanding curriculum, higher teacher expectations, and increased learning, but they did not necessarily attribute those changes to the new standards.

These figures suggest that teachers have largely accepted the new standards, but to what extent have they changed their classroom *practices* in response to the standards? The survey found that 80 percent of the teachers had developed units or lesson plans linked to standards or had modified curriculum to reflect standards.

A more complex picture of teachers' classroom practices emerges from a study of Philadelphia teachers. Elaine Simon and colleagues (1998) found that:

- Most teachers saw standards as a guide for curricular topics, not as a template for designing instructional activities.

- Many teachers stressed that standards were "nothing new" and said they had been teaching to standards all along. Only a third believed they needed to change their instructional methods to align with standards.

- More than 70 percent of the teachers believed that their success in teaching was due to factors beyond their control, particularly student attitudes and abilities. Two-thirds felt that students didn't have the necessary work habits to achieve the new standards; one-third also believed that their students lacked the capability.

If typical, those figures suggest that standards are being absorbed into the system rather than transforming the system. As with many educational reforms, standards-based instruction may bog down when it clashes with well-established teaching norms and classroom routines.

The issue is not whether teachers are willing to be accountable but whether they define accountability in the same way as policymakers do. Charles Abelmann and colleagues (1999) found that teachers framed *accountability* in terms of their responsibility to individual students rather than in conformity to any formal set of rules or guidelines. That responsibility included not just academic learning, but classroom order and student well-being.

Abelmann and Elmore also found that some schools reinforced this individual sense of responsibility with collective schoolwide norms that emphasized particular goals. In still other schools, the collective expectations were reinforced by formal accountability mechanisms (for example, a teacher who didn't fit in might be transferred). Schools defined *accountability* in various ways, but teachers always operated with some inner sense of responsibility. Thus, when states or districts impose a strong *external* accountability system, they never do so in a vacuum. Abelmann and Elmore conclude, "We cannot know how an accountability system will work, nor can we know how to design such a system, unless we know how schools differ in the way they construct responsibility, expectations, and internal accountability."

While we are just beginning to understand how teachers respond to externally imposed standards, we can easily see several areas in which the logic of standards clashes with the culture of teaching.

1. ***Standards-based instruction asks teachers to focus on implementing preselected goals rather than autonomously***

***choosing content.*** In his classic study of teaching, Philip Jackson (1968) found that teachers cherished the ability to choose objectives and activities. Many said that a rigid curriculum or a prescribed set of tightly scripted activities would lead them to look elsewhere for employment. Spontaneity was important. As one teacher put it, "If something interesting comes up, a butterfly flies in the window, we talk about butterflies." The more inflexible the standards, the less appealing they will be to many teachers.

2. ***Standards-based instruction requires heavy reliance on assessments to judge student progress.*** In his study, Jackson noted that teachers gauged their success by immediate behavioral clues rather than by formal assessments. One said:

> I know I'm getting through when the kids are sparking and interested and excited in what they're doing. I think it's the feeling of the class and it's the way the class behaves. I don't think you can tell off in a vacuum, and I don't think you can tell by the objectives, and I don't think you can tell by the tests. It's the degree to which the kids feel part of the activities of the room and participate in them with pleasure.

In general, the teachers Jackson interviewed expressed skepticism about testing, in part because they believed students' performance on tests did not reflect their regular performance. Given a mismatch between tests and everyday performance, many teachers choose daily results.

3. ***Standards-based instruction is based on a highly rational and systematic view of learning.*** In a standards-based system, educators must formally identify essential outcomes, design instruction to achieve them, and use frequent assessment to provide corrective feedback. However, like most other humans, teachers do not always conform to rational models.

Jackson found that the teacher's world was characterized by an uncomplicated view of causality, an intuitive rather than a rational approach to classroom events, an opinionated stance when confronted with alternative teaching methods, and a surface treatment of abstract terms. In short, teachers were not highly logical or analytical in their work, but resembled musicians improvising without a score. Even when successful, they seemed disinclined to analyze their success. "When good fortune strikes, the teachers seem to be saying, it is best not to ask too many questions."

Although this characterization seems unflattering, Jackson speculated that it was a virtue:

> If teachers sought a more thorough understanding of their world, insisted on greater rationality in their actions, were completely open-minded in their consideration of pedagogical choices, and profound in their view of the human condition, they might well receive greater applause from intellectuals, but it is doubtful that they would perform with greater efficiency in the classroom. On the contrary, it is quite possible that such paragons of virtue, if they could be found to exist, would actually have a deuce of a time coping in any sustained way with a class of third graders or a play-yard full of nursery school tots.

4. ***Standards-based instruction assumes that all learners are capable of achieving the goals.*** While most teachers are willing to work hard to take students as far as they can, some harbor doubts about whether the standards are realistic for all students, especially those who are at risk or simply not college-bound. Given the increasing numbers of such students, and the limited time and resources available, the concern is not unreasonable.

Yet the work of Simon and colleagues also suggests that when students struggle, teachers are predisposed to assume the issue is student capability rather than teacher effectiveness, leaving little incentive to invest time and energy in new forms of instruction.

5. ***Standards-based instruction measures progress by comparing students to a standard, not to one another*** (Mary Ann Lachat). Teachers are often desirous that students be judged holistically, taking into account the context of their lives. At grading time, many teachers judge student progress by relative effort and behavior as well as actual achievement (Marzano and Kendall 1997). Standards, by contrast, are unforgiving; students either reach the specified level of mastery or they do not.

6. ***Standards-based instruction transforms teaching from a private to a public activity.*** Teaching has often been charac-

terized as an "isolated" profession in which practitioners spend most of their time interacting with children behind closed doors. Aside from the rare supervisory visit, what happens in those rooms is not viewed by other adults.

In a standards-based classroom, the doors may remain closed, but the results (in the form of test scores) are subject to public scrutiny. Teachers' discomfort is increased by the fact that bare test scores communicate none of the context needed for understanding what actually happened: the at-risk students preoccupied with survival rather than academic achievement; the lack of resources; the frantic pace; and the sheer complexity of directing active young minds toward productive learning.

These observations suggest that integrating standards into the life of a school is a long-term process requiring active reflection, dialogue, and experimentation. Policymakers usually assume that with standards in place, teachers will somehow find the way, but at the district level, this easy assumption is an unaffordable indulgence. Instead, local leaders must put teachers at the heart of the process, eliciting their involvement and hearing their voices (Tell and colleagues 1999).

## Implications for Practitioners

This section explores the implications the new standards have for practitioners, looking particularly at three questions:

How should standards be developed and/or chosen?

What instructional changes will the new standards require?

What support systems must be in place for standards to have an impact?

### Choosing Standards

For many schools, choosing standards seems to be a moot point, since the state has already established some set of learning expectations that will be embedded in statewide assessments. Even when these standards are not explicitly mandated, highly publicized test scores encourage schools to make them a part of the curriculum.

Yet schools that embrace the standards uncritically, jumping immediately to implementation, may shortchange themselves and undermine their long-range success. A standards-based system changes almost everything in a school's life, and unless the standards are understood and accepted, the result will be either a superficial makeover or instructional chaos.

Lauren Resnick and Katherine Nolan (1995) observe that even the best standards documents will accomplish little unless students and teachers adopt them as personal goals. "That will happen only if a concerted effort is made to engage teachers and students in a massive and continuing conversation about what students should learn, what kinds of work they should do, and how well they should be expected to do it."

One compelling reason for a careful review is that state standards are not necessarily classroom ready. Most emerge from a committee process in which diverse views and conflicting agendas are smoothed into a mutually acceptable statement. Agreement may come at the price of vagueness, with standards that are conceptual placeholders rather than clear and concise targets. Marzano and Kendall (1996) state bluntly, "The vast majority of states have standards that are so vague that they will probably have to be reworked—or even totally rewritten—by schools and districts in those states."

Organizations that monitor state standards have frequently echoed these comments. For example, the American Federation of Teachers (1999) rates state standards by a number of criteria, including clarity, specificity, significance, and breadth. By these criteria, the AFT considers only twenty-two states to have clear and specific standards. Similar criticisms (though using different criteria) have been made by the Thomas Fordham Foundation and the Council for Basic Education (Archbold). Although these quality ratings are obviously a matter of judgment and always subject to debate, state standards should not be uncritically accepted as "best practices."

Credibility is another issue. In theory, the term *standard* implies a norm established by people with expertise and used for guiding those less knowledgeable (Ravitch). In many fields, such as engineering, food preparation, and aircraft maintenance, the standards are clear and the need is taken for granted. Experts may occasionally quibble over the details of annual bridge inspections, but everyone expects that there will ultimately be a universal standard supported by technical data as well as professional judgment.

In education, however, standards are almost purely the result of judgment. There are no tests proving that today's children will someday need to know about the Gadsden Purchase or be able to solve quadratic equations; even the most expert educators can only hypothesize what knowledge and skills will be needed twenty years from now. These judgments are by no means arbitrary, but they will vary with a person's background, philosophy, and life experiences.

A Gallup survey of the public found that 72 percent believed high school graduates should "definitely" understand "the causes and global consequences of World War II," but only 6 percent attached similar importance to the standard "understands the growth of states, towns and trades in Sub-Saharan Africa between the 11th and 15th centuries" (Marzano and colleagues 1998). The same survey found that 69 percent agreed that understanding the general nature and uses of mathematics should be required, while only 29 percent felt the same way about understanding of probability. In general, the highest agreement was found in health-related standards, the lowest in the arts.

Some standards are controversial on political grounds. Conservative critics vigorously attacked national history standards as being too "politically correct," and even math standards have come under attack for straying too far from basic skills. Other critics object to the notion that "one size fits all," arguing that diverse students require diverse standards, not uniformity. "We must consider the possibility that the ability to manipulate quadratic equations might not be a realistic goal for all," says writer and former teacher Susan O'Hanion. "Certainly we must not dump kids who don't achieve this goal to the slagheap of high school dropouts" (cited in Gary Stager 1999).

For all these reasons, the initial step in developing a standards-based system is a thorough standards audit, consisting of a series of questions.

1. ***What standards are we required to meet?*** In most cases, this can be determined by reviewing state guidelines, though, as noted earlier, state documents are not always models of clarity. If a standard merely says, "Students will understand and use language arts to communicate," participants may spend considerable time arriving at a common understanding. Fortunately, a growing number of states are providing supporting material such as benchmarks, performance tasks, and lesson plans that should clarify the operational meaning of the standards.

2. ***What standards would we like to meet?*** This is a more complex and far-reaching question, since it taps into each individual's sense of mission and professional aspirations. Teachers, administrators, and community members may never have systematically considered what standards they want students to attain, or they may assume that essential standards are already addressed in the existing curriculum or in state documents.

Districts can use several strategies to stimulate local thinking on standards:

- Review standards developed by national organizations and other states. Mid-Continent Regional Educational Laboratory maintains a compendium of national and state-level standards that can be accessed online (www.mcrel.org).
- Ask teachers to respond to open-ended questions such as, "If you could guarantee that your students would achieve at least one academic goal, what would it be?" Once those responses are on the table, the followup question becomes, "What standards are implied by those goals?"
- Develop a future-oriented educational vision, asking what today's students will need in tomorrow's world (Terry Foriska).
- Ask local employers what standards they believe are most important for students to achieve.
- Ask former students how well their schooling prepared them for work or college. Based on their experience,

what standards are most important for schools to uphold?

The goal here is not to develop unique or wholly new standards. As Marzano and Kendall (1996) note, there is no shortage of thoughtful, well-written standards, and few districts have the resources to match the efforts of state agencies and national organizations. Rather, members of the school community need to take ownership of whatever standards are either required or desired. This can be done only through reflection and dialogue.

Consensus is not necessarily the highest good, however, as the Consortium for Policy Research in Education (1993) points out. Although it would be difficult to implement standards-based instruction with a badly divided school, the goal is to do more than just replicate current practice. If the standards are to drive change, they should, to some degree, take educators out of their normal comfort zones; resistance is likely. One of the leader's key roles is finding the balance between consensus and innovation.

A related question is the public's role in the standard-setting process. Foriska (1998) urges educators to work side by side with community members from the beginning. Marzano and Kendall (1996) agree that public involvement is important but suggest that the nature and timing of their participation should be carefully planned. Formulating standards is a messy, technical process, and involving too many people too early can be "catastrophic." They recommend that the actual writing of the standards be done by qualified educators and then presented to other teachers and community members for advice and feedback.

However it is done, virtually everyone agrees that engagement with the public is important. Stacy Aronson and colleagues (1998) argue that involving diversely representative stakeholders brings two important benefits. First, diversity of viewpoints leads to better decisions as the differing perspectives are reconciled and accommodated. Second, participation builds commitment; as people invest time in the process they begin to feel ownership in it. This ownership will be critical when the standards become inconvenient (for example, when the high-stakes consequences begin to kick in).

3. ***To what degree are the desired standards already being addressed in our curriculum?*** Where standards are concerned, no school starts with a blank slate. There is always a curriculum in place that can serve as a worthy foundation. Standards add precision and accountability to the curriculum, but much of what teachers already do is easily adapted to the new system. Psychologically, it is vital for teachers to recognize that standards do not require them to start over.

One crucial step is mapping out the curriculum in a way that identifies where the desired standards are already being addressed in the curriculum. For example, the Illinois State Board of Education provides schools with a simple form that asks them to list current curricular goals that match state standards (see table 3.1). The Association for Supervision and Curriculum Development has developed a more elaborate mapping process that would also be useful (ASCD).

Addressing standards is not solely a matter of coverage. Many of today's standards go beyond traditional definitions of academic mastery by emphasizing reflection, critical thinking, and metacognition. For that reason, a standards-driven school must also be ready to develop new instructional capacities. For example, the Center for Research on Education, Diversity and Excellence (CREDE) has developed five pedagogical principles that may help students meet the new expectations (Stephanie Dalton 1998):

- Fostering "joint productive activity." The teacher promotes a collaborative environment in which students see themselves as working together on common goals.
- "Developing language and literacy across the curriculum." The teacher accepts and builds on students' existing language patterns and communication styles to develop more sophisticated forms of discourse.
- "Making meaning: connecting school to students' lives." The teacher uses local norms and knowledge to build a bridge to the academic curriculum.
- "Teaching complex thinking." The teacher steadily guides students to more sophisticated kinds of thinking.

**Table 3.1**

## Form Used by Illinois Schools To Match Curricular Goals with State Standards in Science at Early Elementary Level

| STATE GOALS/STANDARDS/ BENCHMARKS | LOCAL GOALS/ OUTCOMES/ OBJECTIVES | 2 | 1 | 0 |
|---|---|---|---|---|
| ***Goal 11**: Understand the processes of scientific inquiry and technological design to investigate questions, conduct experiments and solve problems.* | | | | |
| **A. Know and apply the concepts, principles and processes of scientific inquiry.** | | | | |
| 11.A.1a Describe an observed event. | | | | |
| 11.A.1b Develop questions on scientific topics. | | | | |
| 11.A.1c Collect data for investigations using measuring instruments and technologies. | | | | |
| 11.A.1d Record and store data using available technologies. | | | | |
| 11.A.1e Arrange data into logical patterns and describe the patterns. | | | | |
| 11.A.1f Compare observations of individual and group results. | | | | |
| **B. Know and apply the concepts, principles and processes of technological design.** | | | | |
| 11.B.1a Given a simple design problem, formulate possible solutions. | | | | |
| 11.B.1b Design a device that will be useful in solving the problem. | | | | |
| 11.B.1c Build the device using the materials and tools provided. | | | | |
| 11.B.1d Test the device and record results using given instruments, techniques and measurement methods. | | | | |
| 11.B.1e Report the design of the device, the test process and the results in solving a given problem. | | | | |
| ***Goal 12**: Understand the fundamental concepts, principles and interconnections of the life, physical and earth/space sciences.* | | | | |
| **A. Know and apply concepts that explain how living things function, adapt and change.** | | | | |
| 12.A.1a Identify and describe the component parts of living things (e.g., birds have feathers; people have bones, blood, hair, skin) and their major functions. | | | | |
| 12.A.1b Categorize living organisms using a variety of observable features (e.g., size, color, shape, backbone). | | | | |
| **B. Know and apply concepts that describe how living things interact with each other and with their environment.** | | | | |
| 12.B.1a Describe and compare characteristics of living things in relationship to their environments. | | | | |

*KEY*: 2—Indicates Strong Link; 1—Indicates Moderate Link; 0—Indicates No Link.

*Source:* Illinois State Board of Education, http://www.isbe.state.il.us/ils/science/sclink.pdf

- "Teaching through conversation." The teacher regularly engages students in dialogue that gives full recognition to students' views and values.

CREDE developed these guidelines with an eye to the needs of culturally diverse and at-risk students but considers them applicable to all students. Schools in which teachers had internalized and applied these principles would be much better equipped to help their students achieve the new standards.

4. ***To what degree are we currently achieving the new standards?*** Simply "covering" the standards through curriculum and instruction is not enough; the whole point of accountability is to focus attention on the academic bottom line. Evidence on this question can come from many sources. If the state has developed an assessment keyed to its standards, those test results will obviously provide critical information. If not, districts will need to look to standardized assessments and locally developed assessments.

It is unfortunate that existing assessment tools are not fully aligned with the new standards, so districts will have to use the data in an especially thoughtful way. In addition, community surveys can provide some feedback on public perceptions, though such data do not demonstrate actual achievement. (See chapter 4 for additional information on assessment.)

5. ***Are the chosen standards ambitious, realistic, and consistent with our values?*** If state-level standards are not always good models, how do we know that ours are good enough? Ultimately, only the school community can answer this question, but advice is plentiful.

The Council of Chief State School Officers (1997), after analyzing half a dozen sets of guidelines proposed by various groups, developed six criteria for standards:

- Standards should expect and support all students achieving to high levels.
- Content standards should reflect the strengths of the relevant academic disciplines.
- Content standards should be specific enough to convey clearly the important academic knowledge and skills that all students should learn, but broad enough to allow for multiple ap-

proaches to curriculum, instruction, course design, and assessment.

- Content standards should include a plan for implementation through performance standards, assessments, and accountability measures.
- Content standards should be world-class standards.
- Content standards must be understandable and convincing to the lay public.

Ivor Pritchard offers three criteria: meaningfulness, legitimacy, and practicality. *Meaningful* standards are coherent, understandable, and important, truly reflecting the knowledge and skills that society values. *Legitimate* standards represent a broad community consensus, provide for equitable treatment of all students, and lead to a clear understanding of who is accountable for what. *Practical* standards are understandable to all, coordinated to be realistic in terms of available time for instruction, and coordinated across the curriculum.

Pritchard emphasizes the equity question, pointing out that "standards for all" presents schools with some difficult dilemmas. On the one hand, "in the real and imperfect world, 'all' cannot possibly be absolute." He suggests that a realistic goal would be that 90 percent of future adults achieve at least 90 percent of the standards. On the other hand, exceptions to "all" frequently lead to abuse, particularly when groups of students are exempted by virtue of their ethnicity, socioeconomic status, or life circumstances. Somehow schools must give all students every opportunity to achieve the highest standards, but in a way that does not discourage them

### Changing Instruction

Considering the degree to which standards-based instruction may challenge traditional teaching norms, school leaders can expect little good to come from simply handing teachers a set of standards.

Harris and Carr (1996) encourage schools to start anywhere, observing, "The development of standards-based curriculum is not a linear process. A commitment to standards-based curriculum does not mean that all existing units of study must be scrapped, nor that all units must be designed from scratch."

In particular, they suggest three different starting points. First, teachers can begin by taking existing units and gradually aligning them with the standards. Second, they can begin with students' questions, concerns, and issues, using them to design units that are aligned with the standards. Third, they can begin with essential standards and design appropriate units to support them.

Whatever the starting point, the key is to use standards as a basis for selecting content, planning instructional activities, and developing assessments. Traditional units are often centered around *activities*; that is, teachers concentrate on developing instructional activities related to the topic at hand and tend to judge their success by how well the activity went. Over time, however, teachers can gradually bring each component of the unit into alignment with the relevant standards (Harris and Carr).

Standards mean nothing until they come alive in the classroom—not just in the material the teacher presents, but in the work that the students do. Teachers in a standards-based classroom continually evaluate student performance in reference to the standard, asking questions such as:

What kind of work does the standard call for?

What would a satisfactory product look like?

Does this student's work meet those criteria?

Because many new standards call for complex and sophisticated thinking, evaluating this work will push many teachers beyond long-established evaluative criteria. For this reason, the Annenberg Institute for School Reform recommends that teams of teachers work collaboratively to evaluate samples of student work. Together they can identify the standards addressed by an assignment, create a rough scoring guide, score the work individually, and then seek consensus as a group.

After the work is evaluated, the team can raise essential questions about the implications. What does this work tell us about student knowledge and skills? Was the assignment well designed to achieve the standards? What changes are needed to bring the work up to a higher level of mastery? Over time, teachers can form common criteria for standards-based work (Ruth Mitchell).

In general, standards-based teaching will require teachers to think intensively about assessment, not just about tests, but about a wide spectrum of alternatives such as rubrics and portfolios. Chapter 4 discusses these matters at greater length.

**Supporting the Standards**

Policymakers often explain accountability as providing a "challenge" to students who have not been previously pushed to perform up to their capabilities. But opportunity-to-learn standards remind us that along with the challenge must come *support*.

Even when students have the capability to meet the new standards, they may not be ready to do so immediately (especially for the standards that require complex thinking). When schools ask students to rise to the challenge, practitioners must be ready to provide the support systems that will help students reach the goal. Recall that motivation is highest when a task is matched to the capabilities of the learner. Too little challenge leads to boredom; too much challenge leads to discouragement.

Valerie Lee and colleagues (1999), reporting on their work with school reform in Chicago, argue that both "academic press" and social support are essential. Students exposed to strong academic press *and* strong social support performed best.

Academic press was found in schools with clear, high standards and a consistent "push" for achievement. Social support consisted of personalized, caring environments that respected the individuality of students and encouraged positive, respectful relationships. In some cases, the social support was linked to achievement, as when a school honored good performance by inviting students to have breakfast with the principal, or when high school students were recruited to provide tutoring to elementary students.

Jim Grant and Bob Johnson have identified some barriers to high standards, as well as some possible solutions. One obstacle is the kind of personal baggage that so many children bring to school with them, making achievement of higher standards unlikely. Grant and Johnson recommend a "full-service" approach to education, including close collaboration with social-service agencies.

Another problem is the traditional grade structure that assumes everyone can progress at the same rate. More flexible time options would provide better opportunities to learn.

A third barrier is an overgrown curriculum that tries to be all things to all people, diffusing energy in multiple directions. Grant and Johnson (1997) recommend curricular deflation.

Finally, ideologically driven reforms restrict schools' ability to respond to student needs; a balanced, flexible approach provides better opportunities for more students.

Kate Jamentz provides a useful framework for considering support issues. She notes that a standards-based system requires certain behaviors from each participant, and from the school as a whole. For example, students must be actively engaged in high-quality work, which requires teachers to understand the community's expectations for student performance. These expectations, in turn, require administrators to facilitate communitywide input on standards and ultimately articulate a collective purpose via a clear set of standards. By zeroing in on the bottom line—what is required of students—schools can generate a picture of the support needed.

Finally, Sharon Nelson-Barber (1999) suggests that schools pay special attention to the learning patterns that students bring from home, especially those from culturally diverse backgrounds. She notes that these students often seem the most disadvantaged in a standards-based system, but only because teachers do not recognize that their preferred learning styles may actually be better equipped to handle the new kinds of thinking that today's standards require. For instance, Native-American students often come from communities in which they are encouraged to take responsibility for their own learning, are given choices, and learn through observing and doing. These are exactly the practices that many educational reformers are advocating for all students.

Nelson-Barber argues that the real test of standards will be their success with diverse students. "We must demand that teachers learn to create classroom communities that grant voice and legitimacy to the perspectives and experiences of those who are different from themselves—communities that will not require students to surrender personal and cultural identity in exchange for high academic achievement." To do this, they must work with all students in a variety of cross-cultural settings.

## Standards and Motivation

When policymakers endorse the motivational power of accountability, they are usually thinking of the incentives built into the system. But they also believe that the mere act of setting a standard inspires people to higher performance. As Horace Mann put it many years ago:

> All children, like all men, rise easily to the common level. There, the mass stop; strong minds only ascend higher. But raise the standard, and, by a spontaneous movement, the mass will rise again and reach it.

When carefully articulated and publicized, standards can communicate expectations and inspire achievement. One teacher noted that standards had helped his students focus on the essentials:

> Now I have far [fewer] children who say, "Because [the teacher] wanted me to." That content standard is up there all the time and it gets referred to. (Mid-continent Research for Education and Learning)

Thus, if teachers highlight standards, build them into daily lessons, and incorporate them into classroom assessments, students will take note.

Cognitive attribution theory reminds us, however, that the motivational power of standards depends on a complex interplay of human experience, perceptions, and values. If teachers believe that they are already teaching to standards, or that their students are not capable of achieving those standards, or that this is just one more reform that will fade away, then little will change. The challenge for leaders is not just to put standards at the top of the agenda, but to help teachers see the instructional implications of a standards-based system (see chapter 7).

With standards in place, schools can turn to the issue that arouses more emotions than any other feature of the accountability movement: assessment. Considering the importance that people attach to tests, and recognizing the small but growing number of students who are simply refusing to take high-stakes assessments, practitioners will want to tackle the issue with all the resources they can muster. Chapter 4 explores the uses and abuses of standards-based assessment.

# 4

# Assessment

*In today's accountability systems, tests offer the quickest route to the carrot or the stick.*

"Be a hero: take a zero." Standing on the steps of the local city hall, a handful of Massachusetts high school students waved the sign to signal their opposition to the statewide comprehensive achievement test. Across the state that week, 300 students staged rallies, boycotted the test, or deliberately failed it (Jacques Steinberg 2000). Around the country, a growing number of parents and teachers have joined forces to block statewide tests or at least reduce their influence (Peter Schrag 2000).

Random acts of frustration, or the first stirrings of the next great protest movement? It will be a few years before anyone can answer that question, but the incidents clearly demonstrate the emotional wallop that testing carries. The success of the accountability movement may well depend on the public's willingness to live by the results of the assessments that are a part of it.

Americans have long had a love-hate relationship with testing. On the one hand, they have willingly accepted the claims of psychologists that properly constructed tests can accurately pinpoint knowledge, intelligence, achievement, career direction, and dozens of other characteristics. For example, asked whether teachers who have completed a program with good grades should also have to take a test of knowledge, a staggering 97 percent of the public agreed (Lowell Rose and Alec Gallup 1999).

At the same time, first-hand experience with testing leads many to be wary of the idea that a bubble sheet and number 2 pencil can capture the complexity of the human mind. Thus, test critic Peter Sacks (2000) finds a receptive audience when he cites the stories of Kelly Santos and Gil Madeiros. Santos, the daughter of a teacher and by all accounts an intelligent, above-average student, failed to graduate because she repeatedly tested just below the cutoff point on the math section of the Texas Assessment of Academic Skills. Madeiros, a highly successful attorney and businessman, had to overcome the advice of teachers who interpreted his seventh-grade achievement test as a signal to forget about college and stick to his father's trucking business (Sacks).

Such anecdotes echo the experience of many Americans. One survey found that a majority of parents, while generally supportive of testing, did not believe that assessments were a true and valid measure of their children's abilities (Kathleen Moore 2000).

Thus, policymakers who make testing the linchpin of reform are in a precarious position. On the one hand, virtually everyone takes for granted that testing should be a part of accountability. On the other hand, when the results strike close to home, many are ready to disregard the results as somehow inadmissible.

## The Role of Assessment

The new focus on testing is the result of dissatisfaction with traditional input-based standards of accountability. Because continual upgrades of money, training, and curriculum have had little apparent effect on achievement, reformers have developed an "alternative paradigm":

> Stop fiddling with school resources, practices, services, and regulations, and, instead, state precisely what results are expected. Free schools and educators to generate those results however they think best. In return for that freedom, hold them accountable for producing—and demonstrating that they have produced—the desired results, not for just going through the motions. Reward individuals and institutions

> that succeed. Punish, or intervene in, those that fail (Chester Finn 1995).

The new system requires clear standards *and* a way of measuring whether those standards have been met. "Without a decent means of measurement," says Finn, "the objective, any objective, retains a nebulous, dreamy quality, like getting to heaven, losing weight, or wowing the crowd." Tests are the educational equivalent of thermometers, altimeters, and speedometers; they give us a readout of performance that lets us make crucial adjustments.

Although not everyone shares Finn's enthusiasm about tests, many educators find them helpful as instructional management tools. As Ruth Mitchell (1996) puts it:

> Assessment won't scare you if you remember this important fact: Assessment is information. It is information about what students have learned—regardless of what they have been taught. Standards-based education needs information so that curriculum and instruction can be adjusted. Assessment is your friend.

The role of testing goes beyond simple feedback, however. In today's accountability systems, tests offer the quickest route to the carrot or the stick. Schools and their students will be judged not by how hard they have worked, or how much they have improved, or how well they are doing under difficult circumstances. Rather, the score itself will be the criterion by which rewards and consequences are distributed. Who gets to graduate? Which teachers deserve raises? What schools should be reconstituted? In recent years, "high stakes" and "test" have become almost synonymous, making it difficult for teachers, students, and parents to view the results calmly and dispassionately.

Tests also serve other less obvious purposes. Robert Linn (2000) says policymakers like tests because they are quickly mandated and implemented, and because they provide visible results at relatively low cost. Gary Natriello and Aaron Pallas (1998) add that in a highly decentralized system, tests offer state and federal policymakers an indirect, but effective, means of controlling local schools. And the Office of Technology Assessment (1992) has claimed that testing is often a thinly disguised rhetorical device, designed to confirm preexisting perceptions that schools are not doing well.

## Arguments Against Testing

If testing has consequences, no one should be surprised to find critics among the students, teachers, and parents who are affected. Students worry that four years of school may not lead to a diploma. Teachers face public embarrassment if not loss of income or job security. Parents may find their children's carefully planned future derailed by a poor test score.

But the concerns go beyond simple self-interest. Assessment is a complex process, raising a host of technical and ethical issues. Among the concerns:

1. ***Tests do not adequately capture student achievement.*** When a test uses simple multiple-choice questions, as many do, what gets measured is simple recall, which is far from the kind of complex cognitive skill that the new standards aim at. Eliot Eisner says, "Our children will need to know how to frame problems for themselves, how to formulate plans to address them, how to assess multiple outcomes, how to consider relationships, how to deal with ambiguity, and how to shift purposes in light of new information."

Although a well-crafted multiple-choice question can elicit fairly sophisticated responses (Office of Technology Assessment), most questions rely on memory, allow students to get by with passively selecting rather than actively constructing an answer, and imply that questions always have right answers. No matter how sophisticated, multiple-choice tests "corrupt the teaching and learning process" (Ruth Mitchell, cited by Robert Marzano and John Kendall 1996).

Of course, many of the new state assessments are not standardized, and they are ostensibly tailored to the designated standards. But assessing complex thinking is labor-intensive, driving up costs precipitously. Estimates of increased costs for nontraditional assessments range from three to sixty times the cost of multiple-choice tests (Roy Hardy 1996). Economic pressures will always tend to skew tests toward easily measured outcomes.

2. ***Current assessments have not been adequately validated.*** The concern here is not with validity in a narrow, technical sense; most large-scale assessments easily meet contemporary standards for statistical adequacy. Rather, the problem is one of underlying purpose. A test that is valid for one purpose may be invalid when used for another purpose. For example, W. James Popham (1999) notes that most standardized tests are designed to measure a student's knowledge by comparing it with others' knowledge, *not* to assess educational quality. He says, "Employing standardized achievement tests to ascertain educational quality is like measuring temperature with a teaspoon." In general, few tests have been validated to serve more than one or two of the purposes policymakers have in mind (Paul Barton 1999).

Not only is there often a poor match between the content of tests and the content of the curriculum, the mechanics of the test-construction process tend to eliminate questions that are likely to be answered correctly by most students. Popham points out that items on which students do well reflect the content that the teacher has stressed. "Thus, the better the job that teachers do in teaching important knowledge and/or skills, the less likely it is that there will be items on a standardized achievement test measuring such knowledge and/or skills."

Another validity issue is the match between what gets taught and what gets assessed. This has long been recognized as a problem with standardized tests, which are forced to make certain assumptions about what will get taught in, say, fifth-grade science. A single national test is unlikely to match the curriculum in rural, urban, and suburban schools in Texas, New Jersey, and Georgia. Popham cites a study showing that 50 to 80 percent of standardized test content is not adequately covered in the most commonly used textbooks.

The logic of the accountability movement assumes that assessments *will* be aligned with the standards, but this assumption can break down in two ways. In some cases, states have chosen assessments that fall far short of alignment with the new standards. Perhaps more commonly, the tests reflect the standards well enough, but classroom practice is still aimed at older goals. For example, the assessment may be consistent with standards aimed at higher level thinking, whereas daily lessons are focused more on surface-level knowledge. In theory, poor assessment scores will eventually persuade teachers to change their practice, but in the short run test scores may tell us more about curricular alignment than about instructional effectiveness or student motivation.

3. ***Tests narrow the curriculum.*** Every teacher has faced the same student question: "Will this be on the test?" Every teacher answers carefully, knowing the shrewd calculation behind the question: "If it's not on the test, I can ignore it."

Large-scale assessments can have the same effect. When teachers know their reputation will ride on a certain portion of content, that's where they will concentrate their efforts. A survey of teachers in North Carolina found that they were spending more time on content covered by the test, and were also devoting more instructional time to test preparation (Gail Jones and colleagues 1999).

Of course, this is exactly what policymakers want from a standards-based system; the point of having standards is to identify what is most important. *If* the standards capture the complete vision of what is educationally important, and *if* the assessment adequately reflects the standards, there is no problem. But if the standards focus narrowly on a few areas, or if the assessments are poorly matched with the standards, the curricular agenda may be impoverished. The North Carolina study indicated that science and social studies (which were not tested) were now barely being taught. One teacher reported that her principal did not support the teaching of science and did teacher observations only for test-related lessons.

4. ***Tests encourage cheating.*** When the stakes are high enough, educators are no more resistant to dishonesty than any other group. The rate of overt cheating is not high, but it occurs often enough to be unsettling. In Texas, for example, state officials launched a criminal investigation of alleged record tampering; in Rhode Island, officials delayed statewide assessments because of extensive security breaches; and a Wyoming

principal resigned after alleged test tampering that raised one group's scores from the 42nd to the 87th percentile (Robert Johnston 1999; Jeff Archer 1999; Julie Cort 2000).

Blatant dishonesty is probably rare (or at least rarely reported), but there are more subtle ways of manipulating results, such as excluding students with special needs from the testing, or neglecting to test students who are absent on the official test day. (Statistically, the students who are absent on any given day are the ones most likely to be absent frequently. Predictably, including their scores would lower the results.)

Another morally ambiguous response is teaching students *how* to take tests. Years ago, an old Peanuts cartoon had Linus analyzing the pattern of answers on a true-false test and sagely observing, "If you're smart, you can pass a true-false test without being smart." Linda McNeil (2000) reports that schools are now institutionalizing such strategies, conducting pep rallies with cheers of "Three in a row? No, No, No!" More generally, many schools are explicitly teaching test-taking strategies such as reading the question, using an index card to block out distracting print, and filling out practice bubble sheets (Lucy Calkins and colleagues 1999).

Teachers often defend these practices (with some justice) by arguing that students' futures should not be jeopardized by tests that may be trivial, irrelevant, or inappropriate. But from an accountability standpoint, such practices obscure the information being sought. High-stakes testing may create systemwide a state of affairs analogous to a common classroom problem in which students work to *appear* competent rather than actually become competent (M. Kay Alderman).

5. ***Tests encourage unfair comparisons among districts***. Inevitably, when test results are publicized, citizens eye the scores of neighboring districts as well, sometimes asking sharp questions if their district seems to be lagging. Again, this is fully intended by policymakers, who hope that laggard districts will be shamed into improvement. However, tests do not just measure quality of instruction; they also reflect the abilities and dispositions that students bring with them. Districts with high concentrations of low-socioeconomic-status (SES) students will predictably do worse than districts in more affluent areas.

Comparisons may have the perverse effect of demoralizing the staff rather than encouraging improvement (Heather May 2000).

6. ***Tests are unfair to disadvantaged students.*** The heightened emphasis on testing has spotlighted the longstanding "achievement gap" that shows minority and low-income students consistently scoring below white higher income students on a wide variety of assessments (Robert Johnston and Debra Viadero 2000). For example, Natriello and Pallas examined test results in three states (Texas, New York, and Minnesota) and found that Hispanic and African-American students consistently lagged behind white students. Similarly, Walt Haney found statistically significant differences on the Texas Assessment of Academic Skills that showed a disproportionate number of minorities failing.

Such results do not necessarily mean that the tests themselves are biased; in fact, the cause of the gap has puzzled researchers for years (Johnston and Viadero). However, the disparity does have social and legal consequences. When a single high-stakes test is used as the measure of academic success, minority students are less likely to graduate, with severe consequences for their future (Walt Haney 2000). Not surprisingly, court challenges may follow (the Texas assessment has thus far survived one challenge to its fairness).

In response, advocates of high-stakes testing argue that tests call attention to the discrepancy and will lead to appropriate remediation. Indeed, the other side of the Texas story is that passing rates for African-American students have gone up from 31 percent in 1994 to 67 percent in 1999, while Hispanic passing rates have increased from 39 percent to 72 percent; a few schools have come close to eliminating the gap (Debra Viadero 2000). Some of this improvement is likely due to the fact that Texas schools are held accountable for improving student achievement across several racial categories, not just overall.

Similar questions of fairness arise with students having disabilities or limited English proficiency. What happens when the optimism of "high standards for all" collides with the reality of students who have difficulty reaching traditional expectations, much less today's more rigorous standards? Nationally, no clear answer has emerged. In twelve states, students with disabilities receive a diploma only if they pass the same examination at the same level as other students. In other states, the severity of the handicap is considered in graduation decisions, or special students are exempted from the exam altogether (Lynn Olson 2000b). Some states are considering systems that establish graded diplomas.

Although there is widespread agreement that accommodations should be provided for special-needs students, most states have yet to confront the hard question of whether it is fair to exclude students from the benefits of a diploma because of a score on a test.

7. ***Tests hamstring teachers, robbing them of autonomy and forcing them into a narrow mold.*** As professionals, teachers expect and enjoy the freedom to make judgments, think creatively, and act independently. When assessments are narrowly focused, this autonomy is threatened. In the words of one elementary teacher, "All the things we did that made learning fun and that made children love school are out the window" (Pauline Gough 2000).

Again, many policymakers would argue that this reduction in autonomy is not only tolerable, but desirable. They claim schools are trying to go in too many directions at once, and standards-based assessment applies a necessary discipline to the curriculum. For the good of the students, teachers may have to give up some of their favorite topics or activities. One Washington principal admitted:

> Quite frankly, it [the state-mandated assessment] is not popular because you can't teach a lot of your pet units that you used to enjoy—you know, doing the luau because it was fun... it's a fun unit. We are much, much more directed. (Robin Lake and colleagues 1999)

8. ***Testing takes too much time and distracts students and teachers from learning.*** The headlong rush to accountability may have perverse effects on learning. Barton notes, "There is an impatience at work here that is typically American; it is like pulling up the carrots to see how they are growing."

In Massachusetts, high school students can spend up to 10 days taking components of the statewide test. In many states, the statewide accountability test is not the only game in town; the new assessments have not replaced the older tests, just been laid on top of them. Teacher Joseph Angaran (1999) says,

> In my district, the average 3rd grader is evaluated with the following instruments: a national norm-referenced test for school achievement, a test to determine school ability, and two state-mandated tests in reading and math. In addition, classroom teachers are required to conduct a writing assessment and a test for oral retelling of a narrative or a descriptive text.

Considering that American schools typically have only 180 days maximum in which to do their work, any time taken away from learning is serious. Surveys of teachers indicate that a "huge" amount of time is spent in test preparation beginning weeks before the assessment; half of teachers in one survey said they spent more than thirty hours per year in test preparation (Haney).

9. ***Tests create stress and anxiety among students.*** Classroom teachers frequently report that high-stakes tests create high anxiety and stress for students, and they worry that student self-esteem may suffer. Almost 61 percent of the North Carolina teachers surveyed by Jones and colleagues believed their pupils suffered more anxiety because of the state test, and almost half felt the program diminished students' love of learning. William Mehrens observes that there is little empirical evidence on this point, and cites Ebel's comment that tests are probably less disturbing to children than "angry parents, playground bullies, bad dogs, shots from the doctor, and things that go bump in the night."

But many policymakers would not be disturbed at learning that tests were stressful. The whole point of having a high-stakes test is to create enough anxiety that students will take it seriously. Whether the new state testing programs are actually having a positive motivational effect is still unknown, however, and the effect may differ from place to place. Some researchers believe that a moderate amount of anxiety has positive effects on performance, but that very low or very high anxiety will undermine achievement (Vonda Kiplinger and Robert Linn 1993). It may be the attitudes of teachers, parents, and administrators—rather than the test itself—that determine children's response to assessment.

## Meeting the Assessment Challenge

Clearly, assessment is a minefield for school leaders. On the one hand, meaningful accountability requires some kind of assessment. Without solid information about student performance, schools will never gain credibility from the public and will lack solid data for improving curriculum and instruction. Wrongly handled, however, assessments will mislead stakeholders, generate stubborn resistance from parents, teachers, and students, and derail the educational future of many students. Worst of all, school leaders have to play the hand they were dealt, making the most of tests that have been mandated by higher authorities with little knowledge of local conditions.

The remainder of this chapter synthesizes some basic assessment principles that will help school leaders though the assessment maze.

### Assessment Issues

Anyone who has taken the standard graduate course on tests and measurements knows how quickly the discussion can spin off into arcane statistical analyses. Probably few administrators would lay claim to more than a surface understanding of measurement principles. Fortunately, at a practical level the critical issues can be expressed rather simply.

1. **Content domain.** Every test is aimed at measuring some body of knowledge or skills, from third-grade reading to eleventh-grade science. How that domain is defined, and how questions are extracted from it, will establish the basis of its validity.

Standardized achievement tests, because they are designed to be used across the country, are generic; test-makers analyze commonly used textbooks and curricula to arrive at some middle ground that will appeal to the maximum number of schools. Thus, scores on standardized tests will indicate student performance on basic knowledge and skills that reasonably reflect what "American children" are expected to know and do. However, the scores may be far less accurate in assessing what fifth-graders at Pine Grove Elementary are expected to do.

By contrast, in standards-based assessment the standards *are* the domain. Scanning a set of well-stated standards should allow accurate predictions about what will be on the test. With close alignment, instruction can become much more focused, giving priority to the knowledge and skills that will be assessed.

No matter how a test defines its subject domain, however, it will only supply a sampling of the domain. There simply isn't enough time to measure all relevant knowledge and skills, so test-makers have to create a limited set of items that will accurately reflect the student's mastery of the domain. This calculation always requires some degree of subjective judgment, so every test is assumed to have some degree of measurement error. Because of this inevitable imprecision, assessment experts are unanimous in warning that no single score or test should be used for high-stakes decisions (American Psychological Association 1999; American Educational Research Association 1999; National Research Council 1999).

2. **Basis of comparison.** The interpretation of any test ultimately requires comparing the result to some standard of performance. Traditionally, the most common strategy has been to compare a student's performance with that of other students. For example, a student's score will be reported to be at the 81st percentile, meaning that he or she did as well as or better than 81 percent of other students. This *norm-referenced* standard is simple and easy to understand, and it allows teachers, parents, and policymakers to put student performance in a broad national context.

(It is important to realize, however, that most norm-referenced tests do not directly compare students with their test-taking peers. That is, test-makers base comparisons on a representative group at the time the test is developed. Among other things, this accounts for the "Lake Wobegon effect" in which most schools report their children are scoring "above average." As schools become familiar with the content and format of a test, successive groups of students do better and better. When a new test is introduced, or an old test is renormed, scores tend to dive.)

In addition, norm-referenced tests are inherently competitive; when the slogan is "high standards for all," it can be disconcerting to know that 50 percent of students will, by definition, score below average.

The alternative is the increasingly common *criterion-referenced* assessment, in which scores are compared to a designated level of achievement. For example, everyone scoring above ninety may be considered to be at the "mastery" or "proficiency" level.

The criterion-referenced approach dominates standards-based assessment because it focuses attention on the standards themselves rather than on a comparative ranking of students. Typically, results will say something like "x percent of fourth-graders met the standards at the 'proficiency' level." Setting the "cut score" can be problematic, however, since it always requires some subjective judgment. (How good is good enough?) Unlike the standards themselves, which typically have explicit rationales and have been examined in public forums of one kind or another, cut scores are often determined behind the scenes with little public explanation.

Paul Barton notes that procedures for setting achievement levels are not well established, and Walt Haney has criticized the Texas Assessment of Academic Skills for using inappropriate methods to determine passing levels.

3. **Test format.** The best-known format is multiple-choice, which lends itself to a quick sampling of student knowledge at the expense of cognitive complexity. With the development of intellectually challenging standards, educators have begun to explore testing alternatives that provide feedback on more sophisticated types of knowledge.

- *Performance assessments* seek to emulate real-world contexts in which students can apply the desired skills. For example, students might be asked to demonstrate how they would write a letter of application for a job.

- *Authentic assessments* go one step further by asking students to physically implement the solution. For example, students would be asked to write and actually send a letter of application.

- *Portfolio assessments* ask teachers and students to judge a representative sampling of student work over a period of time.

These alternatives are often better able to capture the more complex kinds of thinking in today's standards, but they require a much greater commitment of resources. They are still relatively rare on statewide assessments.

4. **Scale.** The distinction here is simple. *Large-scale assessments* use a common instrument to measure the performance of large numbers of students (national, state, or district). *Classroom assessments* are the instruments or tasks devised and used by classroom teachers. Because classroom tests vary so much from teacher to teacher, they seldom get much discussion in the accountability debate. But because of their frequency and ubiquity, they can play a major role in integrating standards into the life of schools.

5. **Consequences.** *High-stakes assessments* provide notable consequences for participants. Depending on results, students may fail to graduate, teachers may lose salary bonuses, and schools may face reconstitution. *Low-stakes assessments* provide information, but participants are not directly affected by the results. Both alternatives have advantages and disadvantages. Scores on low-stakes assessments tend not to be distorted by cramming or coaching, but indifference may lead students to perform under their capabilities.

High-stakes tests sharpen motivation, but both students and teachers may focus more on the appearance of mastery than on actual understanding. In addition, high-stakes assessments have moral implications; when a test can lead to denial of diploma or loss of job, test-givers have a deep responsibility to ensure its validity, relevance, and fairness.

**Instruments for Accountability**

As policymakers and educators have attempted to get accountability systems off the ground, a number of specific instruments have helped shape the discussion. Not all these tests will play a direct role in a school's accountability plan, but some knowledge of them is an essential part of "assessment literacy."

1. **NAEP.** The closest thing to a national assessment is the congressionally mandated National Assessment of Educational Progress (NAEP), which since 1969 has provided ongoing assessment in reading, science, mathematics, writing, social studies, and other subjects. Currently, NAEP is a collaboration of the National Assessment Governing Board, an appointed group that sets policy; the National Center for Education Statistics, which administers the program; and contractors who develop the assessments.

Every two years, NAEP samples 120,000 students in grades 4, 8, and 12, focusing on different subjects in different years. To minimize the disruptive effect of the assessment, students are not tested in all subjects, and test sessions are limited to two hours. The goal is to sample achievement in the selected subjects, not to evaluate students, and no names are collected.

The tests are based on content frameworks developed through a consensus process involving teachers, curriculum experts, business representatives, policymakers, and the general public. The frameworks are not standards, but do provide a rough guide to test content. For example, the framework for the 1994 U.S. history assessment indicates that the test will be structured around four themes: change and continuity in American democracy, interactions of cultures, people and ideas, economic and technological changes, and the changing role of America in the world. The framework indicates that the test will assess both historical knowledge and perspective and historical analysis and interpretation. NAEP recognizes three levels of performance: basic (partial mastery of requisite knowledge and skills), proficient (solid academic performance) and advanced (superior performance).

Although dubbed "the nation's report card," NAEP was not designed as an accountability measure but was rather intended to provide a consistent criterion-referenced series of snapshots of student performance over time. As currently conducted, it provides limited evidence on individual students. However, since results are broken down by states, it does offer a simple comparison, and President Clinton proposed using improvement on the NAEP (especially reducing the gap between high- and low-scoring students) as a basis for awarding improvement grants (David Hoff 2000).

2. **National Voluntary Tests.** Thus far the so-called "National Voluntary Tests" exist only as a concept. They represent a carefully hedged bet by federal policymakers who want to be on the record for higher standards but without encroaching on state and local prerogatives. Proposed by President Clinton in 1997, the tests are designed to provide a consistent national framework for assessing student achievement. Unlike the NAEP tests, National Voluntary Tests will report results for individual students, allowing parents, teachers, and others to gauge student progress. As proposed, the tests would assess student achievement in reading at the fourth-grade level and in mathematics at the eighth-grade level, using items very similar to those on NAEP assessments.

3. **TIMSS.** The Third International Mathematics and Science Study was the largest international assessment of students ever attempted. It was administered to students in forty countries in 1994-95, and again (on a more limited basis) in 1998-99 (results for the most recent administration will be released in 2001).

The assessment probed student understanding of math and science on five different grade levels and also examined "contextual factors" such as curricula, textbooks, and student demographics. The content framework for *TIMSS* was developed through a consensus process with math and science educators around the world. The framework includes content (knowledge), performance expectations (using the knowledge), and perspectives (attitudes, habits of mind, careers, and so forth).

As with NAEP, TIMSS distributed its assessment across multiple students. For example, test questions in seventh and eighth grade were divided into eight booklets, but no student completed more than one booklet. Thus it was not designed to assess individual students or schools. However, *TIMSS* has played a major indirect role in the accountability movement by serving as the benchmark for "world class," forcing American educators to define standards in global terms.

4. **Standardized achievement tests.** When people think of testing, standardized achievement tests are the examples most likely to come to mind. For decades, the Iowa Test of Basic Skills, the Metropolitan Achievement Tests, the California Achievement Tests, and similar instruments satisfied educators who felt a need to measure program quality or student progress.

More recently, practitioners, and sometimes the general public, have become disenchanted with the use of these tests, and even advocates of testing are dubious about linking these tests to standards-driven accountability.

Standardized tests offer several advantages for accountability purposes, including relatively low cost and ease of administration. Because they are norm-referenced, schools know where their students stand in comparison to a representative national sample. Even critics such as W. James Popham concede that the tests do a "wonderful" job of comparing student progress to a nationwide sample.

On the other hand, standardized instruments are not always well aligned with most school curricula, and even less so with the new standards. This is especially true when state standards emphasize critical thinking skills.

5. **State assessments.** Virtually all states now require some kind of assessment to be taken statewide. The variety of requirements makes it difficult to generalize; one assessment consultant says, "The one constant is that they are all different" (Lynn Olson 1999a). Forty-two states use criterion-referenced tests, often aligned with state standards, but twenty-nine use some form of standardized test (meaning that some states require both). While multiple-choice questions are still common, many states have incorporated performance items into the assessments; two have made portfolio assessment a part of the process. English and math are the most frequently tested subjects.

The quality of state assessment systems appears to be variable. A 1997 study by the advocacy group FairTest criticized many states for overreliance on norm-referenced multiple-choice tests that were poorly aligned with standards. The group further criticized some states for testing too often, for failing to review the effectiveness of their assessment systems, and, most of all, for using a single test as a requirement for graduation.

Some of these problems may simply be birth pains as states move into the assessment business and begin to sort their way through the complexities of measuring student performance. Over a period of several years, the American Federation of Teachers has found noticeable improvement in state standards and assessment systems, and many state-developed instruments will likely continue to improve as their makers gain experience.

6. **Classroom assessments.** Teacher-made tests get relatively little attention in the accountability arena, undoubtedly because their diversity defies generalization. Lacking common content, and administered under a variety of conditions, classroom tests provide no basis for consistent measurement of student performance. However, they can play a vital role in moving schools toward greater alignment with state standards. When classroom assessments begin to present students with the same kind of cognitive challenges as the standards, the new forms of thinking become an everyday activity, not just an annual event.

## Developing an Assessment System

Clearly the stakes in high-stakes testing go beyond determining who graduates or gets a bonus; the effectiveness of the entire accountability system is on the line. Without accurate knowledge of results, schools will remain rudderless in their efforts to reach higher standards.

Assessment makes everyone edgy and somewhat defensive, with the specter of public embarrassment never far away. School leaders can expect to deal with a full range of emotions from members of the school community: anxiety, frustration, anger, and despair, interspersed with moments of relief and, on rare occasion, joy. Keeping a steady course and a level head is a major leadership challenge.

Worst of all, state-mandated systems leave local leaders with limited control and few options, sometimes having to live with an inflexible and poorly designed system that no one at the local level asked for. But even under those circumstances, principals and district officials can take affirmative steps to ensure that assessment is carried out in a way that results in maximum benefit to teachers, students, and parents. In the long run, principals who are both knowledgeable and passionate about the role of assessment can help others understand the benefits and limits of assessment, and can even hope to have some influence on testing practices.

Because assessment policies vary so much from state to state, detailed action plans are impractical in a book intended for a national audience. However, the guidelines that follow provide the nucleus of an informed and thoughtful approach to the assessment labyrinth.

1. ***Be clear on purpose.*** Policymakers have multiple reasons for mandating assessments, not all of which are compatible or achievable with the same instrument. Lorraine McDonnell (1994) interviewed federal and state policymakers about their hopes for assessment, and found at least seven distinct purposes:

- providing information about the status of the educational system
- aiding in instructional decisions about individual students
- bringing greater curricular coherence
- motivating students to perform better and parents to demand better performance
- acting as a lever to change instructional content and strategies
- holding schools and educators accountable for student performance
- certifying individual students as having attained specified levels of mastery

An assessment that is effective for one purpose will not necessarily be appropriate for other purposes. One of the defining acts of leadership is pointing a direction when everyone is confused. Leaders can serve their schools well by asking (and helping answer) the crucial questions: Why are we giving this test? How will students, teachers, and parents benefit? What will we learn from the test and why do we need to know it? Even when the state-mandated assessment appears to be badly designed or poorly aligned with standards, leaders can help the community understand the limits on what can be learned.

2. ***Evaluate existing tests.*** Once the purpose is clear, schools can evaluate their overall testing program. For each test, the key questions are:

- What does this test accomplish for us?
- Does it give us meaningful feedback on our goals?
- Is the test valid for the purpose we have in mind?
- Do we actually use the information from this test to make decisions about placement, promotion, remediation, or curriculum?

A careful evaluation may discover that the national achievement test given every two years has a very poor match with the district curriculum. Or, it may turn out that scores from the test are simply filed and forgotten.

3. ***Align assessment with the standards.*** The essence of standards-based accountability is a tight link between standards and assessment. If the content of the assessment does not reflect the standards, then the effort will be wasted, or even counterproductive as the immediacy of the test pulls teachers' attention away from the standards. Without careful alignment, accountability becomes a buzzword devoid of substance.

In districts coping with mandated state assessments, the alignment has ostensibly occurred at the state level, and the district's major responsibility is making sure the standards are reflected in the curriculum. (However, some state tests are not well aligned, presenting districts with the awkward situation of trying to boost student achievement on an assessment that has little to do with the standards.)

One area in which the school does control alignment is classroom testing. Robert Marzano and John Kendall (1996) argue that classroom assessments—not national or state tests—should be the primary tools for evaluating standards-based accountability. For one thing, teachers know students better than anyone else in the system. In addition, large-scale assessments usually measure only a small sampling of student performance, making them prone to measurement error. (Marzano and Kendall note that the measurement error on a typical standardized achievement test may range up to an entire year in grade-equivalency scores.)

Classroom assessments, while less elaborate than large-scale assessments, will cover more ground over the course of a year, reducing the impact of measurement error. Finally, aligning classroom assessments with standards begins to focus instruction on the desired goals. No matter how strongly the new standards have been injected into the curriculum, they will have little impact on learning unless they also show up on weekly tests and quizzes.

Classroom assessment does not change easily. The standards often involve more than just new content; they also require more complex forms of thinking that will require new kinds of test questions. Professional development aimed at test-writing strategies may be necessary. Marzano and Kendall point out that teachers will need a diverse array of assessment tools, not only the traditional forced-choice items (true-false, multiple-choice) but essays, performance tests, and portfolios.

4. ***Use multiple forms of assessment.*** If testing experts agree on one thing, it's that high-stakes decisions should never rest on a single measure (American Psychological Association 1999; American Educational Research Association 1999; National Research Council 1999; Learning First Alliance 2001; Monte Neill 1997). Using a variety of assessment tools provides a more accurate readout of student capabilities and gives more students a chance to perform under optimum conditions. Performance assessment, portfolios, self-evaluation, and exhibitions (for example, senior projects) should be a part of the overall assessment system.

5. ***Make fairness an integral part of the system.*** High-stakes testing has powerful ethical implications. To deny a diploma is to change the course of a life—not the kind of thing that should be done lightly. The testing disparity between minority and white students is an unmistakable red flag to anyone concerned about students' futures, as is the assumption that the new standards apply even to those with disabilities. Justice requires both that all students be held accountable to the standards *and* that all students have (1) an opportunity to learn the standards and (2) a chance to express their knowledge in the best way they can.

Concrete answers here are elusive; as noted earlier, the achievement gap has thus far resisted all efforts to erase it. Partial solutions may be found in using a variety of assessments; allowing appropriate accommodations; being sensitive to language differences; providing remediation for those who fail high-stakes tests; and making sure that all students have access to adequate resources and opportunity to learn (American Educational Research Association).

6. ***Practice openness.*** If a school has any dirty laundry, it will be flapping in full view when test results are published. The potential for embarrassment is discomforting but inevitable, so school leaders should set the right tone by putting aside defensiveness and addressing the results openly and honestly. Communication should be steady before, during, and after the assessment.

Parents and other community members should hear the full story: what efforts the school is making to meet the new standards; what the assessment is like and what strengths and weaknesses it has; what the results are and how those compare with previous tests; and, above all, what the assessment has taught us and how it will help us improve instruction next year. Successes should be recognized and applauded, but too much fanfare this year may backfire next year when scores take a dip. Assessment is a tool for improvement, not an athletic contest.

7. ***Treat assessment as a learning opportunity.*** This goes to the heart of the accountability paradigm: Assessment is the corrective steering that keeps the whole enterprise on track. Ide-

ally, people should deal with test scores as dispassionately as a thermostat deals with a change in temperature. We know they don't, of course. High stakes keep everyone on edge; students worry about their future and teachers worry about their reputation, while the school's critics eagerly wait for an opportunity to pounce on a poor showing. However, the school's leader can provide a welcome steadying influence by calmly and persistently asking, "What have we learned?"

## Assessment and Motivation

Does assessment motivate students and teachers? Any educator knows the unambiguous answer to that question: yes! The real question is *what* it motivates them to do.

In accountability theory, assessments are the crucial bridge between goals and consequences. They offer tangible, systematic evidence that teachers and students are fulfilling expectations, thereby triggering the appropriate rewards or sanctions. Knowing this, students and teachers will take the necessary steps to ensure that the standards are met.

Stated that way, the theory is simple and compelling, but actually somewhat misleading. When policymakers decree that achievement of standards will be measured by high-stakes tests, they ensure that *motivation will be focused on passing the test, not on achieving the standards.* The distinction is subtle, but significant.

When students and teachers are focused on standards, we can expect them to engage in continual study, inquiry, and reflection as they seek answers to crucial questions: What is the goal? How close are we to achieving it? What steps would bring us closer? In other words, the standards function as guides to the learning process, encouraging thoughts and actions that lead students closer to the goal.

By contrast, tests require students only to demonstrate that the standards have been achieved (or, at the minimum, to create the *impression* that the standards have been achieved). Test evaluators see only what the student has put on paper, not the activities and thought processes that have preceded the test; as long as the answer falls within predetermined limits, the student will get credit for achieving the standard. For purposes of earning rewards or avoiding sanctions, a correct guess is as good as a genuine insight.

Thus students may find it advantageous to rely on shrewd guesswork, while teachers may be tempted to improve student guessing by teaching test-taking skills. (Ironically, a strong test-preparation effort may actually diminish the instructional time available for teaching the standards themselves.) Teachers may also decide to teach to the test, narrowly focusing instruction on concepts likely to be tested. More rarely—but still too often—both students and teachers may cross the line to overt cheating. In short, high-stakes testing unquestionably energizes students and teachers, but not necessarily in the way that policymakers intend.

In addition, high stakes can distract and demoralize students. Most motivational specialists say that *moderate* anxiety improves performance but *high* anxiety undermines it. When a lot is riding on the outcome, worry about one's performance can inhibit performance (Moshe Zeidner 1998). A simple thought experiment illustrates the point. If you lay a four-inch-wide beam on the ground, you can probably walk it end-to-end without difficulty. Now imagine that the same beam stretches from the roof of one skyscraper to another.

Finally, high-stakes assessment can diminish teachers' sense of autonomy. In theory, today's accountability promises teachers freedom of methods in exchange for achievement of designated standards. But knowing that their reputation, their compensation, and their students' future depend on the outcome of a single test, most teachers will see themselves as having only a narrow range of instructional options. Under those conditions, they are unlikely to have much enthusiasm for standards.

None of these motivational considerations eliminates the need for assessment, which is essential to keeping the system on track. But they do support the growing number of educators who have argued that accountability should not rest on a single high-stakes measure. Although the logic of accountability requires assessment, it does not require that rewards and sanctions flow from a one-time, all-or-nothing test. Measuring performance over time, using a variety of methods, lowers the stakes on any given occasion and makes it more likely that both students and teachers will focus on achieving the standards, not just passing the test.

# 5

# Use of Performance Data: Indicators and School Report Cards

*Will widespread public knowledge of schools' performance lead to demands for improvement?*

Americans who used to wait with sweaty palms for grades to be passed out probably take a certain amount of satisfaction in knowing that educators themselves now receive report cards and go through some of the same anxious moments.

Today's accountability demands that test scores as well as other "quality indicators" be heavily publicized. The assumption is that since the school is a publicly funded, democratically governed institution, widespread public knowledge of performance will lead to demands for improvement. Moreover, when results are public, professional pride will lead educators to work even harder to achieve good results. As the Southern Regional Education Board (2000) put it:

> Ratings raise awareness, provide focus and energize schools and communities to work to improve student achievement. At their best, ratings can provide momentum, measure schools' progress and show parents, the public and policymakers that schools can improve.

Or as a participant in an earlier SREB study said, "What gets measured gets taught. What gets reported gets taught twice as well" (Southern Regional Education Board).

Early experience with school report cards has shown, however, that merely publicizing results does not by itself lead to clear improvement strategies. Karen Levesque and colleagues (1996) note that school leaders may find the data useful for public-relations purposes, but that teachers do not find them useful in assessing their own performance or that of their students. Numbers are just numbers, and, without a meaningful context in which to interpret them, they may confuse as easily as clarify. Just as important, schools must be prepared to *use* the data to generate focused improvement strategies.

This chapter examines the role of performance data in supporting school improvement. The first section describes a variety of educational indicators that can reveal the school's "vital signs." The second section tells how selected indicators can be packaged and presented in a "school report card." The chapter closes by discussing the ways that schools can go beyond mere reporting to become truly data-driven, using results to guide reform.

## Educational Indicators

An educational indicator is a statistic that shows something about the performance or health of the education system. For example, test scores or dropout rates are often believed to reveal how well a school is doing. However, mere bits of data do not constitute indicators; the figures must also have some reference point or standard of comparison to make them useful. To say that a school's fourth-graders achieved an average score of 83 is meaningless unless we know how well that satisfies the designated standards or how it compares with other schools (Jeannie Oakes 1986).

Usually, an indicator will fulfill one of the following purposes:

- Describe the school's performance in light of some accepted benchmark.
- Describe features that are associated with desired outcomes (for example, instructional time).
- Illuminate central features of the school that are usually a sign of overall health (for example, curriculum offerings).

- Provide information relevant to current policy or problems (for example, a school concerned about disparate test results would be interested in data showing how different ethnic groups fared on the statewide test).

What indicators *cannot* do is determine the significance of the information or decide what steps should be taken as a result. School performance is not like economic performance, in which everything can be reduced to a common scale of value; human judgment is always required (Richard Shavelson and colleagues 1991a). Oakes adds this caution:

> Indicators can provide valuable information to guide the debate and dialogue about whether school is sick or well, effective or ineffective. But judgments about the health of the educational system can only be made by interpreting indicator data in the context of educational values and experience with schooling.

This can be seen by imagining that a school's fourth-graders performed at the following levels on a state mathematics assessment:

"Advanced"—7 percent

"Proficient"—29 percent

"Basic"—37 percent

"Needs work"—27 percent

What should we make of these results? Is this about what we could expect? What actions should we take as a result? Answers will be elusive until we ask some additional questions:

- How does this compare with other fourth-graders around the state?
- How does this compare with the score of last year's fourth-graders at this school?
- How does this compare with the performance of this group when they were third-graders? (In other words, how much gain has occurred?)
- To what degree have students had the opportunity to learn the material? For example, have standards been integrated into the curriculum?

- To what degree do demographic factors affect the scores?

Although we can eventually gain confidence that we know what the scores mean, we are still faced with the question of what action to take. Should the priority be getting students from "needs work" to "basic," or is it more important to increase the numbers attaining "proficient"? And what should we be doing differently to get that kind of improvement? Do we need to adopt new instructional approaches? Hire better qualified teachers? Reduce class size? Whatever the answers to those questions, it is not the indicators that will tell us, but our own professional judgment. The indicators merely ensure that our judgment is grounded in the current reality.

Beyond these issues, Oakes suggests that good indicators have several technical qualities:

- They measure features common to many schools, thus permitting meaningful comparisons.
- They measure enduring features, not transient events.
- They are readily understood by stakeholders.
- They can be gathered in a way that is relatively convenient and economical.
- They are generally accepted as valid and reliable statistics.

## Types of Indicators

Schools have a multitude of statistics to choose from (as anyone responsible for filling out state and federal reports can attest). What should be included?

According to Oakes, "to properly specify which indicators should be a part of a system, we need a model of how the education system works." Thus, if we believe that teacher quality plays a major role in student achievement, we need some way to measure teacher quality. Unfortunately, there is no single model of educational success, which means that indicator systems can vary dramatically from state to state and district to district.

Shavelson and colleagues (1991b) note that indicators fall into three broad categories: those that measure inputs, such as money; those that measure processes, such as school governance; and those that measure outputs, such as test scores. The current accountability movement emphasizes outputs, but inputs and processes may also be worth measuring because they constitute enabling conditions that lead to outcomes.

Robert Linn and Eva Baker (1998) classify indicators by the degree to which schools can control them. One category contains indicators that schools are held directly accountable for, such as test scores and attendance. A second category contains items that are only marginally under the control of schools, such as the percentage of teachers with appropriate certification. (Although schools are responsible for hiring, they have no way of ensuring that supply will meet demand.) The third category contains indicators that are beyond the control of schools, such as student demographics. Linn and Baker argue that it makes most sense for schools to focus on what they can control.

States and districts use a wide variety of indicators in reporting results:

1. *Test scores.* These are almost universally included, because everyone agrees that student achievement is the bottom-line indicator of school health and effectiveness. Although not everyone agrees that existing tests are an adequate measure of achievement, in today's accountability climate their presence is unavoidable. However, great care is required in presenting and interpreting the information.

2. *Other achievement indicators.* For those who believe that tests fail to tap the full range of student achievement, other measures exist. Gerald Bracey (2000) lists the following: percentage of students taking college-admissions tests, percentage of students taking advanced-placement courses or International Baccalaureate tests, percentage of students going on to postsecondary institutions, and student success rate in college.

3. *Student behavior.* These types of indicators include data on attendance, school persistence, and disciplinary actions. While these do not directly predict achievement, virtually ev-

eryone assumes that orderly classrooms and student attendance provide crucial enabling conditions for achievement.

4. *Instructional quality*. Indicators may focus on curriculum content, class size, and instructional methods.

5. *Teacher quality*. This category includes data on the number of teachers teaching out of certification or major, number of substitutes employed, degrees earned, years of experience, and percentage of teachers involved in professional-development activities.

6. *Resource base*. While arguments persist over the link between student achievement and size of budget, most people take the commonsense view that there is *some* relationship between resources and achievement. Possible indicators include the size of the budget, the percentage allocated to instruction, adequacy of facilities, and age of textbooks.

7. *Stakeholder satisfaction*. In terms of consequences to schools, the real bottom-line indicator may be the degree to which the community is satisfied. Measures of satisfaction from different stakeholders—parents, businesses, higher education institutions—can prove enlightening, and some districts are even experimenting with student and parent evaluation of teachers (Jay Mathews 2000). Bracey suggests that the degree of parental involvement is another good indicator.

8. *Contextual factors*. Some information, while not directly related to performance, helps to illuminate performance data. The most obvious example is demographic information; test scores will mean very different things in a suburban district where 85 percent of high-school graduates go on to college and in an urban district with high concentrations of students from households with low socioeconomic status.

## Deciding Which Indicators To Use

Administrators who wish to create an indicator system should exercise selectivity. Indiscriminate data collection not only costs time and effort, it overwhelms those who need to use the data, mixing the significant with the irrelevant. Moreover, the mere existence of data can incite action that is misguided. If a test score is low, the immediate impulse is to act to raise it—regardless of the reasons it was low to begin with. Even when a particular indicator conveys potentially valuable information, it can have perverse policy effects.

For example, some states are now tracking performance data by racial/ethnic groups. While such tracking has the benefit of making disparities obvious, it could, in one scenario, lead to abandonment of the assessment as discriminatory, rather than prompting efforts to improve performance. (In another scenario, it could lead educators working with high-need populations to shrug and say, "You can't expect too much.")

When choosing indicators, board members and educators should be able to answer three questions:

- Why is this information important?
- How much effort is required to track the data?
- How will we use this information when we get it?

## Preparing the School Report Card

School report cards originated in the comparative "wall charts" developed during the Reagan Administration by Secretary of Education Terrell Bell. Bell's purpose was to put pressure on an education system that he felt was badly lagging in performance.

Although the wall charts were much criticized as superficial and misleading, the underlying concept has thrived as state after state has mandated some form of public reporting of school performance; by 2001 forty-one states are expected to have them. The content and formats of the reports are as diverse as the states that created them. Some just contain numbers with no commentary, whereas others run to a dozen pages or more (Lynn Olson 1999b).

Nor is their effectiveness proven. A survey by *Education Week* found that a majority of educational stakeholders had never even seen a school report card, nor had most of them sought out the information. Whereas 90 percent of community members thought that such public reports would motivate teachers to work harder, few were able to explain how they themselves could use the information as a lever to improve schools; many parents said that poor performance would simply drive them to transfer their children to different schools (Olson 1999d).

Districts typically have little to say about state report cards, whose contents are mandated by legislation. Once the data are transmitted to the state, the education agency compiles the information and distributes it according to the designated format. (Examples are available on many state education agency websites.)

States sometimes ask districts to create and distribute their own report cards, and even when this is not the case, districts may find it to their advantage to formulate their own report—if only in self-defense. A growing number of companies and nonprofit organizations are setting up websites that provide objective data as well as more subjective ratings; some of them, in the words of a California principal, "are just awful."

Schools that publish their own performance reports can establish the proper context for understanding the information. And by tailoring the report to suit the needs and interests of a local audience, they can make themselves the first and most important stop for anyone seeking information on the school. This section explores possible strategies for designing such a local report card.

## Developing an Interpretive Context

As noted earlier, naked facts and figures can create confusion and sometimes misperceptions. Thus, the first step in developing an effective report card is to decide on an interpretive context that will make the information meaningful. This requires not just gathering data, but asking the questions that will actually result in change. Why are we highlighting these particular data? What do they tell us about the health of our schools? How can we use this information to make changes that will bring us closer to our goals?

**Purpose.** For policymakers, the primary purpose of public reporting is to satisfy accountability, but stakeholders have more practical needs that go beyond that abstract principle. Everyone in the educational system has specific kinds of questions that require solid answers. What kind of an education is my child receiving? Is the school providing a humane, child-friendly environment? What are the most urgent educational and developmental needs of our students? How can we improve student achievement?

Unless we examine data that can answer such questions, the reporting system will serve little practical purpose. Karen Levesque and colleagues (1998) note, "If performance indicators are not rooted in goals, the indicators themselves may become the de facto goals, and improving the data—rather than the underlying performance—may become an end in itself."

Goals can come from many sources. For example, a well-crafted mission statement formulated with broad involvement of the school community will provide useful guidance on district priorities. Some goals may be found in strategic plans; others may surface in discussions of the local inservice committee. Whatever the source, Levesque and colleagues (1998) suggest that helpful goals have six characteristics:

- meaningful (having a clear meaning and apparent value)
- realistic (being achievable in a reasonable time)
- complementary (supporting other key goals)
- prioritized (allowing choices to be made when resources are limited)
- agreed to (accepted by stakeholders)
- measurable (permitting a specific level of performance or numerical target)

**Theory.** As noted earlier, data make sense only in the light of some theory or model of education. Raw information by itself provides little direction. If test scores are down, we can agree that is not good, but that does not tell us what steps to take. Rather, we first consult our implicit theories about what it takes to raise achievement—a more tightly focused curriculum, for example, or qualified teachers, or reduced class size.

These theories tell us the kind of information we should be seeking. If we believe class size is critical to achievement, we need to collect data on class size. If we think that time on task is a critical variable, we need to generate data on how teachers are using class time. If we believe that orderly classrooms are the key to learning, then we should be collecting information on student behavior.

Without focused information that allows us to act on our theories, test scores alone are likely to generate random behavior. (Our theories may prove to be wrong, of course, but the right kind of information at least allows us to realize it and redirect our efforts.)

**Audience.** Not everyone in the educational community has the same information needs. Parents want to know the kind of experience their children are having; staff members need to know where to focus their improvement efforts; community members may just want general reassurance that things are on the right track.

When *Education Week* conducted focus groups with parents and citizens, test scores were considered important, but not the most important; school safety and teacher qualifications topped the list for parents. Parents tended to share teachers' concerns, such as class size, but were more attracted to performance items than to input measures such as spending.

While agreeing that test scores were important, most stakeholders had some skepticism about the value of the available information; a majority felt that test scores should not be the only item by which schools should be judged. For example, parents were equally interested in school safety, teacher qualifications, and class size. In addition, many community members were not greatly interested in data on the school's demographic context, in part because they felt this information was being offered as an excuse for the schools' inadequate performance (Lynn Olson 1999b).

**Narrative.** To be an effective communication tool, performance reports should tell a story. As with any good story, the school's report will have several critical elements.

***"Here are the goals we set."*** Every story is in some way a quest; the dramatic tension comes from knowing that the protagonists are trying to achieve some worthy goal and that they will encounter obstacles along the way. The stated goal may simply be improvement over last year's score on the state assessment, or it may be aimed at a more specific target (for example, "at least 80 percent of seventh-graders will achieve at least the 'basic' level on the state math assessment"). In other schools, the goal could be increasing attendance or reducing disciplinary incidents.

***"Here's how we tried to reach those goals."*** This is often a missing piece of the story; parents and other community members usually see the year-to-year results but are not always aware of the efforts that schools have made to improve those outcomes. This part of the report gives schools a chance to let the public see that the intervening year has not just been more of the same old methods, but that focused initiatives have attempted to address district needs and boost student achievement.

***"Here's how we did."*** This, of course, remains the heart of the report, an inescapable bottom line that forces everyone to confront reality. The goals were achieved, or they weren't.

***"Here's what we think the results mean."*** This is the beginning of the crucial next step to put the data to work. If the scores fall short of the goal, what are the likely causes? If we did achieve the goals, what accounts for that? Which of our efforts seemed to have an impact? Results do *not* always speak for themselves, and careful interpretation may be needed (see next section).

***"Here's how we're responding."*** Whatever the results, they should generate new goals for the coming year. Based on our interpretation of what happened, we should be ready to enhance the initiatives that seemed to work and drop or adapt the approaches that seemed unproductive.

**Interpretation.** Almost half a century ago, Darrell Huff (1954) made many Americans conscious of something they had probably sensed for a long time when he said the best way to lie is with numbers: "A well-wrapped statistic is better than Hitler's 'big lie'; it misleads, yet it cannot be pinned on you." But deliberate deception is less of a problem than the "spin" that interested parties give the numbers. Few people understand the statistics behind a set of test scores, so they tend to take the numbers at face value and fit them into whatever conceptual framework makes the most sense. For some, the glass ends up half full; for others, it's clearly half empty.

Educators walk a fine line in providing an interpretation in their performance reports. On the one hand, raw numbers can be mystifying or even misleading, so some kind of interpretive context will help the public make sense of the data. On the other hand, interpretation that appears defensive, evasive, or self-serving will probably be dismissed by a public with a low tolerance for spin doctoring.

Interpretation begins with an explanation of how the numbers are generated, including some idea of the nature of the assessment. (Parents who assume the math assessment is about basic computation may rethink their interpretations when they see sample questions that ask students to solve a life-like problem and then explain why their answer is correct.) For example, Coronado Village Elementary School (1998) in California says this of their district writing assessment:

> A District developed Writing Assessment has been given to students in grades 2 and 5 for three years. The goal of this assessment is to improve the quality of student writing and thinking across the curriculum. With training on how to score the tests, teachers used a rating scale (or rubric) to score the papers. A three-point rubric was used for grade 2, while grade 5 was scored using a six-point rubric. Teachers were asked to review individual results with each student and discuss ways to improve the writing.

Second, there should be some indication of what level of performance the numbers represent. Is a score of 45 considered basic, proficient, or advanced? In turn, what do those labels mean in practical terms? (What can "advanced" students do that "proficient" students cannot?) For example, the New York State report card (New York Department of Education 2000), which shows four levels of achievement on the fourth-grade mathematics assessment, explains level 2 this way:

> These students need extra help to meet the standards and pass the Regents examination. They show some knowledge and skill for each key idea for elementary students, but no

> knowledge and skills for the key ideas for intermediate students. They use basic mathematics facts, work with whole numbers, and identify units of measurement. They can use manipulatives to solve for an unknown.

Third, what is the trend line? How do this year's scores match up against last year's? If we aren't at the desired level, are we at least heading in the right direction? The answer may not be immediately clear, since year-to-year changes in performance can occur for any number of reasons. For example, any veteran teacher can testify to the fact some classes have unusual collections of talent, resulting in stellar performance throughout their school careers, whereas others struggle for everything they get. A "star" group (or the converse) can give a misleading impression about current-year scores. In addition, simple measurement error may be responsible for other changes. Karen Levesque and colleagues (1998) warn, "Be cautious about concluding that a trend exists when you have only a few data points and the changes are small." Only when the shift persists for several years does it constitute a trend.

No matter how uncomfortable educators are with the idea of being compared with other schools, the public will inevitably do some kind of benchmarking against the performance of other schools, either nationally or locally. (Local news media routinely publish side-by-side comparisons of schools in their circulation area.) For that reason, performance reports should provide information that lets people make valid comparisons.

The first requirement is making sure that comparison is apples to apples, something not necessarily guaranteed, and not always easy to determine. For example, Oakes notes that "attendance" sometimes means the number physically in attendance, but at other times includes those who have "excused" absences.

Second, the community should have some idea of how meaningful school-to-school differences are. Even a few points may loom large unless people understand that the difference is within the range of measurement error.

Finally, consumers of the report should understand the contextual factors that affect relative performance. A school with high numbers of students who are low in English proficiency, low in socioeconomic status, or transient will be at an obvious disadvantage when compared to an affluent suburban school.

This contextual information needs to be handled deftly, however. As noted earlier, the public is likely to see hand-wringing about demographic factors as a kind of excuse-making. In addition, continual harping on student background may not only offend groups in the community but may inadvertently reinforce existing stereotypes about certain kinds of student who "just can't learn." The slogan "high standards for all students" may be unrealistic in the short-term, but it constitutes a moral commitment that schools cannot walk away from.

One way out of this dilemma is to adopt a "value-added" approach to assessment reporting. Value-added assessment, developed by William Sanders (1998) and used as the foundation of Tennessee's accountability program, attempts to measure longitudinal student growth rather than taking a cross-section of performance at a particular grade level. Thus, instead of simply noting whether this year's fourth-graders did better than last year's fourth-graders, value-added assessment measures how much this year's crop of fourth-graders have learned in the past year.

The value-added approach is fairly new, and not yet universally accepted among measurement specialists, but it provides a commonsense way of getting around demographic variables. Unfortunately, the statistical procedures needed to implement it are complex, and may be beyond the resources of schools or smaller districts. Statewide networks or consortia could provide the necessary expertise (Dennie Palmer Wolf and Ann Marie White 2000).

## Presenting and Disseminating the Report

One of the challenges of putting together a school-performance report is that not everyone wants the same level of information. Too little information, and some people will become frustrated with the incomplete picture. Too much information, and people are overwhelmed. There is no perfect solution, though some districts have used a multilevel approach, providing a kind of "executive summary" for those who are just interested in the basic information, with more detailed supplements for those who want more. Web-based reports readily lend themselves to this approach, with a main page of basic information containing hyperlinks to more detailed data.

Another issue is readability. A report jammed with figures presented in no apparent order and with little accompanying narrative is as inviting as a telephone book. An uncluttered, visually appealing layout will make it more likely that community members will actually read the document. For example, the Los Angeles Unified School District has adopted a standard graphic modeled after a dashboard (see figure 5.1). Skillful use of visuals such as bar graphs, pie charts, and frequency distributions can also enhance the report's appearance and readability.

Concern for a clean appearance should never get in the way of communicating essential information. Readers often need clear explanations of what the data represent (for example, "percent" and "percentile" are often confused).

Once the report is prepared, there are many options for disseminating it. It can be presented at PTA meetings or annual district meetings, printed in the local paper, mailed to parents and key community stakeholders, or placed in libraries, doctors' offices, and banks. Putting the report on the district website will make it available for inspection at any time during the year.

Communication need not end with dissemination of the report. Kate Jamentz (1998a) has described an approach called "accountability dialogues" in which stakeholders come together to achieve common understanding on standards, performance, and accountability. The emphasis in these sessions is not on "educating" the public, but on mutual listening that gives community members and educators a chance to express their concerns and simultaneously understand the concerns of others. Jamentz notes that such "authentic accountability" is not easy, requiring "not only courage, but capacities for public outreach, data collection and analysis, effective communications, group

**Figure 5.1**

## School Performance Report Used by Los Angeles Unified School District

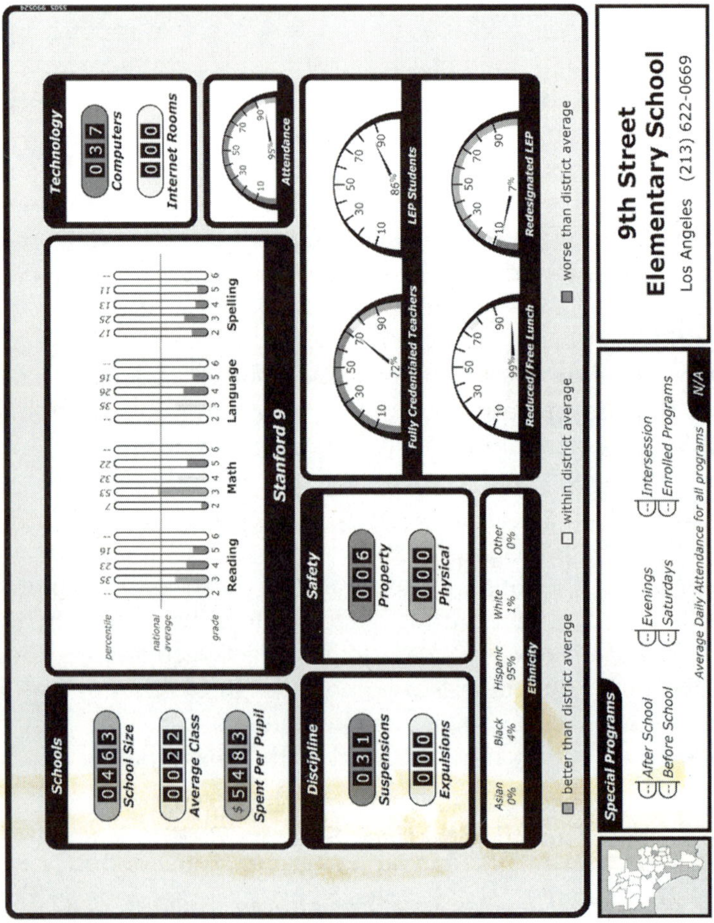


*Source:* Regents of the University of California. *School Report Card*. Copyright © 1999 Regents of the University of California. Reprinted with permission.

process and conflict resolution—all beyond what is typically demanded by daily life in schools." The payoff for these efforts is the restoration of public trust.

## Using Data To Improve Performance

For the public, simply knowing the results may be sufficient. If the school's performance is good, parents and citizens can turn their attention elsewhere. If the performance is poor, they can complain to policymakers, seek a different school for their children, or simply shake their heads and vote against the next levy increase.

For educators, however, the data trigger a much more challenging process. With concrete evidence of their successes and failures, educators face the expectation that something will be done to improve performance. Good information allows schools to become data-driven, focusing their energies and avoiding scatter-shot efforts to reform.

The Annenberg Institute for School Reform has described data-driven schools this way:

> As an integral part of their practice, schools with effective accountability systems examine their practices, explicitly, publicly, and collectively.
>
> Each school is guided by an ethic of continuous improvement so that it is proactive, willing to make changes based on that data. Continuous improvement becomes the school culture. Questioning, seeking data to inform those questions, reflection and action are simply "the way things are done."

In one sense, data-driven schools are just an extension of good professional practice. Sharon Rallis and Margaret MacMullen point out that many teachers already exercise this kind of internal credibility. "Put simply, practitioners who are accountable evaluate their own practice and then use the information to improve." They also note that the transition from this kind of everyday accountability to systematic use of data is often hesitant because the information mandated by the state is not what teachers are looking for. For example, many teachers feel that standardized tests do not capture the full reality of a student's performance. Thus, the leader's challenge is to find ways to weave the externally mandated data into the daily professional lives of teachers.

To ground the collection and use of data on improvement of performance, the Annenberg group has outlined a six-phase process they call the "inquiry cycle" (Lorraine Keeney):

1. ***Establish desired outcomes.*** This is the key that allows one to make sense of the data and to determine where to go with it. If we can't answer the question, "What do we want our students to be able to know and do?" then the data will be so much noise. This can be a lengthy and taxing process, requiring considerable debate and dialogue to achieve a workable consensus, but it is an essential starting point.

2. ***Define the questions.*** These are more specific questions about the school. To what degree do our children learn those things we consider most important? What's happening at school that we like? That we don't like? Where can improvements best be made? Some of these questions may come from parents, taxpayers, or state policymakers; others will arise from the daily life of teachers.

3. ***Collect and organize data.*** Before running out and generating reams of data, participants should ask some focusing questions. What kinds of evidence would best answer our questions? Where are these data located (or how can they be generated)? How can they be collected in a timely and economical way?

Given the quantity of data available, schools can easily be overwhelmed at first; narrowing the scope with some well-chosen questions keeps the task manageable.

Levesque and colleagues (1998) suggest several practical considerations as the data are collected. First, decide how often data will be collected. In some cases, such as state assessment results, the information becomes available once a year at a predetermined time, but in other cases the frequency of data collection is for the school to decide. Decisions on frequency may affect the quality of the information. Surveying teachers on instructional practices once a year may result in less accurate information; surveying them weekly would become burdensome.

Second, decide where and how the data will be stored. The key information may be scattered throughout multiple databases. It may or may not be worth the effort to consolidate the key information, but it should at least be tagged so it can be pulled together efficiently at the appropriate time.

Third, determine who will be responsible for collecting and analyzing the necessary data. This is especially important when the desired information is not already a part of existing databases.

4. ***Make meaning of the data.*** The numbers alone are meaningless; the raw data have to be transformed into patterned information that will assist in making decisions. For example, if we are concerned about reading ability of students, knowing the average score for the school's fourth-graders is of little use until we tease additional meaning from it with the right questions. How does this score compare with those of other schools having our characteristics? Within the overall score, are there areas or particular skills that are lower or higher? Do these scores relate to curricular patterns? Are there racial/ethnic group differences in the scores?

This analysis in turn may lead to additional questions focused on *why* scores are at this level. Maryland's department of education has suggested a number of clarifying questions that might be important. Do staff members understand what students are asked to know and do on the assessments? Do staff members know how to teach the content standards and performance outcomes? Are the outcomes tested included in the curriculum taught in your school? Do staff members use the outcomes in everyday instruction? What school-improvement strategies did we implement last year that we hope would impact these data? Did the strategies work? (The complete list of questions can be found on the department's website, www.mdk12.org.)

In analyzing the data, it is often tempting to converge quickly on an explanation, especially if our improvement strategies went smoothly and performance has gone up. Levesque and colleagues (1998) remind us, however, that "linking cause and effect is difficult to do, especially in a school setting where a complex mix of inputs and practices influence student outcomes." They advise considering all possible explanations for the results, as well as carefully examining the improvement process to ensure that it actually went as planned. (For example, have teachers truly changed their instructional practices or just renamed them?) Before a strategy is enshrined as the solution, this explanation should be supported by multiple forms of data that can stand close scrutiny.

5. ***Take action.*** A careful analysis of the data should suggest some kind of action. If there is a curricular hole, it should be plugged. If teachers are not regularly monitoring student performance on the standards, classroom tests can be reconfigured to measure the desired outcomes. If last year's improvement strategy has made a difference, it should be enhanced and extended.

Kate Jamentz, noting the importance of teachers' sense of efficacy, has linked this action step to teachers' beliefs that they can have an impact:

> We can use our look at data to build that sense of efficacy. How are we bringing data to the table so that it ends in action? We shouldn't sit around and ask questions of data that don't end with something like "Okay, what should we try tomorrow that might make a difference." (Cited in Keeney 1998)

6. ***Assess and evaluate actions.*** Actions are "best guesses"; some will hit the target, others won't. The cycle begins again.

As described by Keeney, this six-step process is driven by a dialogue between the ideal (expressed in the mission and goals) and the real (expressed in the data). The data can be uncomfortable, but facing them uncompromisingly can lead us closer to the ideal.

At Roberto Clemente Middle School in Philadelphia, students ranked at the bottom of the city's middle schools, lived in poverty, and had many behavior problems; about a third of the teachers left every year. The faculty, wanting to examine the high rate of behavioral problems, established a database that could sort problems by teacher, student, type of behavior, and number of offenses. What they discovered was that students were coming to school without knowledge of how to behave properly. Faculty reported that some teachers had opposed providing students with constant supervision, but the data convinced them that increased supervision could significantly reduce infractions. Within two months, inappropriate behavior was reduced over 95 percent (Keeney).

Although Annenberg's inquiry cycle logically begins with goal setting, as a practical matter it can begin anywhere. For example, data can be collected without having established a clear purpose. While this is less efficient, some schools have reported that "swimming around in the data" for a while allows issues and questions to gradually float to the surface where they can be acted on.

In the end, the sequence of the inquiry cycle is probably less important than the underlying processes that support it: collaboration, willingness to examine the results candidly, reflection in developing strategies, and a commitment to go where the data point. The challenge for leaders is to use information designed for external accountability and to weave it into the fabric of teachers' everyday lives.

## Public Reporting and Motivation

Open access to information is a hallmark of democratic institutions, and even before the onset of the current accountability movement, school test scores were commonly published in local papers. But accountability advocates argue that brightening the spotlight on key indicators will add motivational fuel to the school-improvement process.

First, educators will feel the pressure to maintain a positive image, especially when results are accompanied by comparative ratings. Few teachers want to be publicly branded by working in a school ranked as "deficient" while their crosstown rival is judged "successful." Some policymakers are blunt about their desire to "shame schools into success."

Second, parents and community members will put pressure on their schools to improve unacceptable results, or may even vote with their feet by transferring their children to other schools.

Although there is little question that public reporting grabs attention, its motivational effects are still unproved. Educators who want to avoid embarrassment may be tempted to manage public perceptions rather than improve student performance. At least in the short term, explaining away poor results is easier than boosting scores.

In addition, schools with limited resources and large populations of disadvantaged students may find themselves repeatedly labeled "inadequate" or "in crisis," even after unstinting efforts to improve. Teachers in struggling schools already know their students are not getting the education they should; publicly rubbing it in may only create despair and defeatism.

The public response to school report cards is even less clear. Early evidence, such as *Education Week's* surveys and focus groups, indicates many citizens have not seen their district's performance report and cannot describe what steps they would take to press for improvement. Rather than lobbying for better results, many parents suggest they would be more likely to seek another school for their children. Although loss of students would be a powerful sanction for underperforming schools, we know little about how many parents would actually follow through with such a threat. Nor can we be sure that standards would be the driving factor in student withdrawals; many parents define *school quality* broadly, showing as much interest in school safety and class size as in test scores.

Thus, while public reporting of results is likely to keep attention centered on standards, it will not in all cases generate the kind of sharply focused behavior that policymakers desire.

# 6

# Incentives: Motivating by Rewards and Sanctions

*The growing use of rewards and sanctions puts teeth into the new accountability systems.*

Teachers have always carried the mantle of "responsibility," taking it for granted that they have obligations to their students and, in a more abstract way, to the public. But that sense of responsibility has always been tempered by the hard reality that thoughtful teaching does not necessarily lead to meaningful learning. When students are poorly motivated, weighed down by poor home environments or absorbed by the pleasant distractions of popular culture, teachers' best efforts may be fruitless. For that reason, teachers usually define professional responsibility as working long hours and trying one's best.

Today's accountability, however, demands *results,* and, if the results are not forthcoming, consequences. Outside the profession, most people take for granted that good teachers should be rewarded and poor teachers should be penalized, and they scorn the traditional salary schedule that purportedly rewards mediocrity by paying for longevity and course-taking. Chester Finn and Danielle Wilcox (2000) state flatly that "greatness of teaching... is not in the eye of the beholder. It's in the hard evidence of how much and how well one's pupils learn."

Admittedly, action steps have lagged behind the rhetoric; most accountability advocates have inverted Teddy Roosevelt's advice by talking loudly and wielding a rather modest stick.

Policymakers have floated ideas such as reconstituting or closing low-performing schools, tying teacher compensation to student performance, and even dismissing principals and teachers in low-performing schools, but implementation of such ideas has been spotty (Lynn Olson 1999c).

Nonetheless, states and districts are inching toward a new way of doing business:

- Florida gives vouchers to students in schools that have earned an "F" rating for three straight years (Jessica Sandham).
- In Aurora, Illinois, teachers with classes that exceed a 25 percent student-failure rate are required to develop a specific plan to reduce the number (Tracy Dell'Angela 2000).
- Cincinnati teachers will progress through a five-stage pay schedule from "apprentice" to "accomplished," based on achievement of specific teaching goals (Julie Blair 2000).
- In Chicago, third-, sixth-, and eighth-graders must achieve a specified score on an achievement test before they will be promoted (Melissa Roderick and colleagues 1999).
- Dallas high school students can earn $100 for each Advanced Placement test that they pass (Stacy Teicher 2000).

Clearly, states and districts are now willing at least to experiment with incentive systems focused on student achievement. This chapter describes the rewards and sanctions being used to put teeth into the new accountability systems. The first section briefly reviews the rationale for incentives and discusses some key implementation issues. The next two parts describe incentive systems for students and teachers, respectively. The final section looks at state takeovers and school reconstitution.

## The Power of Incentives

The theory behind incentives is based on a behaviorist principle that coincides with a common-sense assumption:

Behavior that is rewarded will be strengthened. Much of what people do is driven by incentives. We shop at a particular store because the advertisement promised a bargain rate; we show up at work regularly because we want the monthly paycheck; and we tune in a particular TV show because we anticipate good entertainment. We even buy lottery tickets—a bet that no sensible person would take—because the potentially huge payoff makes the modest investment seem reasonable.

As logical as the theory seems, however, it is often forgotten in the day-to-day turmoil of organizational life. Richard McKenzie and Dwight Lee (1998) note that this is especially true in large organizations, where it is difficult to get everyone moving toward the same goals. The people at the top often have clear objectives and thoughtful strategies but lack the technical expertise and street-level knowledge of the employees who must do the actual implementation. For example, principals may have a clear understanding of state standards and an indepth knowledge of curriculum, but they must rely on teachers to tailor the material to the needs and capabilities of the particular students in their classrooms.

Savvy leaders delegate many decisions downward, but the greater the freedom provided to employees, the greater the risk that they may not act in the company's interests. McKenzie and Lee note that failure to perform desired actions is not a sign of laziness, selfishness, or perversity. Rather, everyone comes to work with a personal agenda that does not necessarily match the organization's priorities.

Incentives can help overcome these difficulties in two ways. First, they clearly communicate the desired goals. Organizational leaders say a lot of things, some sincere, some not, some just irrelevant. But when the organization backs the talk with tangible incentives, it sends a clear message about priorities. Second, incentives make it worthwhile for employees to put those goals on their personal agendas. Ideally, then, organizations can allow maximum freedom of action to employees on the front lines while letting the incentives pull behavior in the right direction.

McKenzie and Lee are strong advocates of incentives, but they caution that poorly designed reward systems can have perverse effects. They cite the case of secretaries whose typing output was measured by automatic keystroke counters, and who boosted their output by spending lunch hours typing unnecessary documents. Similarly, computer programmers who were paid by the number of bugs they identified often deliberately created the bugs they then "fixed." Nonetheless, McKenzie and Lee state without reservation, "Incentives work and always have, often with dramatic effect."

Incentives work only if they are linked to some clear measure of performance: "If you produce *that*, you will get *this*" (David Osborne and Diane Plastrik 2000). Osborne and Plastrik identify five different kinds of results:

- quantity (increasing the number of outputs)
- efficiency (reducing the cost of outputs)
- quality (improving timeliness, responsiveness, or accessibility)
- impact (improving the effectiveness of the organization's work)
- cost-effectiveness (improving the ratio between cost and effectiveness)

In the current school-accountability climate, the key measure is impact—specifically, improving the level of student achievement. Choosing the actual instrument or means to attain that impact, however, calls for careful thinking.

Osborne and Plastrik offer several principles to guide the choice:

1. Any measure that will determine incentives should be objective rather than subjective. Subjectivity creates suspicion that rewards will be based on personal relationships or bias rather than on actual accomplishment. (Traditionally, teachers' immediate reaction to merit-pay proposals is fear of unfairness.)

2. The measured outcomes should be ones that employees can affect. No matter how attractive the incentive, it has little value if workers believe the goal is beyond their control. Osborne and Plastrik cite Michael Quinn Patton's formula:

Demand to produce outcomes

- control over outcomes

+ high stakes

---

= corruption of indicators

3. "Be careful what you target—you might get it." The need for measurable, objective outcomes sometimes drives leaders to choose targets that are convenient but impoverished. If the district uses an off-the-shelf standardized test because it is cheap and easily administered, then teachers will focus on whatever knowledge and skills are represented on the test. Usually, these are only pale imitations of the cognitive complexity of the new standards.

4. Because incentives tend to drive out behavior that does not lead to the targeted outcomes, leaders should consider attaching incentives to a range of behaviors and goals. Thus, even though student achievement is the ultimate goal in schools, incentives can also be attached to *processes* (such as adopting a new instructional strategy).

Linking incentives to processes gives rise to several issues. Teachers tend to favor the idea, because they can control processes more easily than they can influence student achievement. However, the connections between process and product are not as clear in teaching as they are in other fields. It is often difficult to say exactly what strategies will result in higher achievement for a particular group of students. At the very least, if incentives are attached to processes, they should be based on strong empirical research that shows a link between process and product.

5. "Very small sums can go a very long way with people." Organizations can get a lot of mileage out of "psychic pay" that offers recognition and appreciation. Examples include plaques, award ceremonies, certificates of achievement, news stories, dinner with the CEO, and lapel pins. These small gestures may be especially effective in schools, partly because differential compensation is still controversial and partly because teachers themselves are oriented toward psychic rewards (Dan Lortie).

6. Organizations should strive for an incentive system that balances positive incentives with negative consequences. While emphasizing the importance of avoiding a "climate of fear," Osborne and Plastrik also argue that negative consequences get the attention of employees who might be willing to forego the more positive incentives. Possible consequences include negative publicity from failure to meet goals, loss of privileges or autonomy, intervention from above, or loss of income or job.

## Incentives for Students

Students occupy a complex role in the social ecology of schooling. They are the clients around whose needs the system revolves, yet unlike clients in other fields, who can freely choose to seek or reject help, students have moral obligations and are expected to work hard and do as they are told. Because learning can only happen with their cooperation, they are also held accountable.

As numerous analysts have observed, incentives for students have been weak at best. Hard work may bring derision from peers yet have little apparent payoff in practical terms. Surveys show that while businesses often require diplomas, they pay little attention to grades and typically don't even ask to see transcripts (Meg Sommerfeld 1995).

Moreover, despite universal admonitions to work hard and stay in school, the culture itself does not enthusiastically support learning. "Cognitive achievement or mastery, whether narrowly academic (scoring well on tests, getting high grades) or broadly intellectual (engaging voluntarily in the world of ideas), is not widely valued in America" (Arthur Powell 1996). American parents seek a "balanced" regimen for their children and are likely to give as much emphasis to soccer as to homework. While the existing system seems to provide some built-in incentives with an elaborate structure of tests, grades, and diplomas, many students discover that those hurdles can be overcome with half-hearted effort or even ignored outright.

For that reason, consequences for students form an important part of the accountability landscape. The possibilities seem to fall into five major categories:

- making promotion contingent on performance
- making graduation contingent on performance
- providing extrinsic rewards for student performance
- providing intrinsic rewards for performance
- linking achievement to employment

**Promotion Policies**

If any idea has come to symbolize the new accountability, it is the battle cry "no more social promotion!" The idea of ill-educated students sliding through the system year after year stands as the essence of everything that accountability is designed to change.

For all the rhetoric on social promotion, hard data have been scarce. The U.S. Department of Education (1999), in the midst of a major initiative against social promotion, had to admit:

> Some states do not collect retention information at all and many others collect only limited data. Retention rates vary widely and it is difficult to interpret and compare the data. Social promotion remains a hidden problem.

To marshal evidence against the practice, the DOE was forced to rely on highly indirect indicators such as surveys showing the numbers of graduates who could not balance a checkbook.

Nonetheless, there is little doubt that many students can progress from grade to grade without having to demonstrate mastery of specific academic competency. The American Federation of Teachers surveyed eighty-five urban school districts in thirty-two states and found little consistency and considerable vagueness in promotion policies. District policies often said little more than that promotion depended on "sufficient growth in learning basic skills." Actual decisions were based on an ill-defined mixture of standardized test scores, grades, attendance, and teacher recommendations. Teachers themselves often report passing students who have not achieved as expected.

While accountability advocates clearly favor an end to social promotion, simply using retention as a consequence is not necessarily effective. For one thing, research has failed to show that retention improves student achievement (Center for Policy Research in Education; United States Department of Education 1999). As numerous critics have pointed out, there is little reason to expect that students who have struggled to understand material for an entire year will benefit from simply repeating the same methods the next year.

And, while most parents support an end to social promotion, their philosophical support may fall away when their own children's progress is affected. Anyone who has worked in schools knows that parental pressure sometimes plays a role in educators' decisions about student advancement.

Especially in the early stages of standards-based reform, when large numbers of students may fail to meet the standards, wholesale retention is not viable. In Washington State, for example, 80 percent of fourth-, seventh-, and tenth-graders failed to meet designated standards in reading, writing, listening, and math (Keith Ervin 2000). Simply retaining those students would be administratively difficult and politically impossible.

The complexities of promotion policies can be seen in Chicago, which has been conducting a closely watched effort to end social promotion (Roderick and colleagues). Students in the third, sixth, and eighth grades must attain a target score on the Iowa Test of Basic Skills to be promoted. Those who fail to achieve the target on the first round of testing in May are placed into the "Summer Bridge" program for extra instruction, and they are retested in August. Those who still fail to achieve the criterion are retained. (Waivers are available for students with special needs or LEP background; almost a third of third-graders were exempted in the first year of the program.)

After three years, the Chicago policy was showing mixed results. Overall, many more students (especially sixth- and eighth-graders) were now meeting the identified standard; students with the lowest skills showed the greatest improvement (from a 12 percent to a 49 percent pass rate for eighth-graders, for example). However, students who managed to qualify through the summer program did not show much improvement the following year, suggesting that the initial improvement may have been a testing artifact. Moreover, students who were retained continued to struggle; only a quarter to a third made "normal" progress the following year.

The Chicago researchers, considering these data, drew several conclusions:

1. The combination of additional instruction and threat of retention resulted in improvement for many students, especially in the higher grades. Third-graders seemed to be less affected by the threat of retention or less able to engage in the concentrated learning provided by the Summer Bridge. As a consequence, Chicago is adding the summer program (without retention) for at-risk first- and second-graders.

2. The retention policy has not addressed the adequacy of the instructional program during the school year. What didn't work in the first year of third grade won't necessarily work in a second year.

3. Chicago's progress is attributable to two factors. First, the district has allowed considerable flexibility in implementation, thus defusing potential resistance in the beginning. Second, the district has supported the program with added instructional time and reduced class sizes. The authors caution, "Do not attempt to implement this policy unless your school district is willing to invest, as Chicago has, substantial fiscal and administrative resources."

## Graduation Requirements

Graduation is an even more likely place to target consequences, since it serves as a major cultural as well as academic milestone. Graduation is a highly visible and highly anticipated event for high school students and their families—a signal that childhood has ended and a new phase of life is under way. In addition, the diploma is a major qualification for employment, and few people question that it is a key to further schooling or decent employment.

By the beginning of 2001, only eighteen states required passing a standards-based assessment for graduation (Lynn Olson 2001). Despite the general public acceptance of accountability, policymakers have encountered strong opposition to test-linked graduation standards, and some states have delayed or scaled back exit testing.

For example, Wisconsin had to settle for a watered-down requirement that allows students to compensate for low test scores with good grades, teacher recommendations, or other factors to be decided by local boards. Students may opt out of the test altogether with parental permission (Julie Blair 1999). The legislature backed off a much stricter requirement because of strong opposition from parents (including the state PTA) who either didn't see a need for the test or were concerned about attaching high-stakes consequences to a single test.

Another issue is fairness. Because the high school diploma serves as a social sorting mechanism, students who fail to earn one are effectively condemned to the lower tiers of the economy. For that reason, policymakers must provide for students with special needs, students with limited English skills, and students who simply don't perform well on paper-and-pencil tests.

One policy option is "differentiated diplomas," which tries to avoid an all-or-nothing approach by offering "certificates of attendance" for those who fail to meet the standards or, conversely, honors diplomas for those who have met the standards to a high degree. Although almost half the states are experimenting with such distinctions, little is known about how these differentiated diplomas are regarded by colleges, employers, and the public at large (Lynn Olson 2000b).

### Extrinsic Rewards

Instead of threatening sanctions, some districts have chosen to provide a variety of positive rewards for achievement—or sometimes for just showing up for the state assessment. Among the incentives are pizza parties, laptop computers, free parking, sports tickets, exemption from final exams, college scholarships, savings bonds, and plain old cash (Bess Keller). While critics denounce such practices as bribery and claim they take the attention off meaningful learning, some administrators say they at least send the message that academic performance is valued; the tangible rewards have an immediacy that adult admonitions seldom achieve.

### Intrinsic Rewards

Some critics have suggested that student motivation is best improved by offering them a positive school experience. Powell suggests that the school context itself can be an incentive. Purposeful, well-focused school communities with a clear sense of mission can be appealing to many students. Personalization, with adults who know and care about students on an individual level, can also be a powerful incentive.

### Employment Decisions

Although businesses frequently complain about the quality of high school graduates, their hiring practices have not always taken advantage of the student-performance indicators that might distinguish good workers from poor workers. In recent years, the Business Coalition for Education Reform (BCER) has encouraged its members to ask prospective employees for data such as grades, attendance rates, and disciplinary records, and then to use that information in hiring decisions. Business groups have also worked with schools to develop transcripts that are clear and easily understood by future employers.

## Incentives for Teachers

The existing teacher-compensation system was not designed for accountability. Since the early decades of the twentieth century, teachers have been governed by the single salary schedule that bases pay on a combination of experience and education. Any teacher with three years of experience and a bachelor's degree will be paid exactly the same as any other teacher with those qualifications. The system works well to ensure equity and objectivity, but it does not differentiate levels of on-the-job performance (or, as critics would say, "it rewards mediocrity").

Over the years, the single-schedule system has proved remarkably resistant to change. Periodically, districts have initiated what they call "merit pay," but most such efforts are either short-lived or have little connection to merit (Allan Odden and Carolyn Kelley 1997). In some cases, the plans fail because they operate competitively, offering rewards to a relatively small number of teachers. Sometimes the selection criteria are vague, failing to communicate exactly how teachers can qualify. In other cases, the criteria have little relationship to results and are merely extra pay for extra work. Finally, merit plans have not always been well funded; after a round or two of implementation, budget pressures divert the money elsewhere. (Alternatively, the merit awards may be so small that they fail to be motivational.)

In the 1980s, policymakers were briefly enchanted by another form of merit pay, "career ladders." Ladders were designed to give master teachers an "upward career trajectory" by publicly recognizing their skill, assigning them leadership responsibilities, and increasing their pay. However, the result was often to take these teachers out of the classroom, so the increased salary ended up as a reward for nonteaching duty. In addition, the number of teacher-leader slots was limited, allowing only a relatively small number of teachers to be involved. Finally, as with merit plans, career-ladder funding often proved to be unstable (Odden and Kelley).

Teachers themselves seem to be mostly concerned about the subjectivity in many merit proposals. The profession does not have a universal definition of good teaching, and teaching seems to involve many intangibles that are difficult to measure. One teacher said:

> At my school, I would be hard-pressed to point out someone who I think is not working hard for the students. Do I think they are all equally effective? No. How to evaluate that without creating massive discontent? Very difficult. (Lewis Solmon and Michael Podgursky 2000)

A related concern is bias and favoritism. Solmon and Podgursky quote another teacher who said:

> Performance-based compensation for teachers would only strengthen the good old boy network. Since the talented teachers are often considered to be the ones rocking the boat, they are not usually members of the good old boy club. Therefore they would not be the teachers earning the performance-based compensation.

## Alternative Compensation Systems

Despite the checkered history of merit proposals, the accountability movement of the 1990s has again generated proposals to pay teachers for what they produce. Odden and Kelley have suggested that the information-age economy demands new forms of compensation.

Historically, organizations (in the private sector as well as in schools) based salaries on *jobs*. That is, they agreed to pay employees for performing a set of predetermined duties that required a particular set of qualifications. More exacting duties and higher qualifications brought higher salaries. In recent years, however, many businesses have found that jobs evolve too rapidly for that kind of approach. Instead, organizations are shifting to a competency-based approach that compensates employees for specific *skills* that may be applied across a variety of jobs, or in a series of jobs.

Teaching is changing in similar ways. Odden and Kelley note, "Teachers teach, counsel, plan, manage, develop curriculum, train colleagues, evaluate practice, develop budgets, monitor progress, run meetings—in short, engage in a wide range of 'jobs,' each of which requires expertise to perform well." Conceivably, teaching may be ready to move beyond job-based compensation.

What are the alternatives? Odden and Kelley suggest four possibilities:

- Competency-based pay rewards employees for demonstrating specific skills that are applicable to their work.

- Performance-based pay rewards employees for achieving targeted outcomes that are valuable to the organization.
- "Gainsharing" rewards employees for developing processes that improve efficiency or effectiveness.
- Contingency pay rewards employees for engaging in certain activities, such as training programs.

Although Odden and Kelley see all these options as applicable to teaching, they suggest that competency-based and performance-based systems hold out the best opportunities for meaningful change.

## Paying for Competencies

The traditional salary schedule indirectly recognizes competence by paying teachers for course credits, but degrees and credits are at best "uneven in quality, general in nature, and probably not specific to school needs." The newer proposals are much more specific about the skills and knowledge that are being rewarded.

Odden (1997) describes three broad areas of competence:

1. *Classroom instructional competencies.* To be effective, teachers need a firm grasp of content, competence in employing an array of instructional strategies, and a good understanding of curriculum. This is especially true with the implementation of more rigorous standards for student learning; teachers who may be highly adept at teaching students algorithmic approaches to mathematics may be less prepared to lead them to think critically and independently about problem-solving.

2. *Other educational tasks.* Beyond classroom instruction, today's teachers may be called upon to expand their expertise to new areas, serve as counselors, develop curriculum, or devise assessments. These supplementary skills are especially valuable as schools undertake major restructuring.

3. *School management and leadership.* School reform requires active involvement of the entire staff, and schools can benefit by having a sizable number of faculty members with specific skills to run meetings, construct and implement budgets, and evaluate programs.

Odden and Kelley argue that a competency-based pay system would allow districts to encourage and reward the specific skills needed to implement local reform. In addition, a competency-based system would offer teachers a rich career-development pathway. In a profession often characterized as having a flat career trajectory, skill-based compensation would allow teachers to define their careers in terms of growing expertise rather than progression through hierarchical levels.

Teachers could demonstrate skills in a variety of ways. An increasingly popular route is to pass the rigorous field-based assessment of the National Board of Professional Teaching Standards. Some states and school districts have offered teachers significant bonuses or pay increases for board certification, ranging from a one-time $2,000 award to $10,000 a year for five years. In addition, any time teachers acquire added certification or endorsements (especially in an area of shortage), extra compensation could be provided. Finally, Odden and Kelley recommend developing skill assessments based on criteria such as the standards of the Interstate New Teacher Assessment and Support Consortium. These standards reflect a broad consensus of the skills needed by beginning teachers, but could be extended to higher developmental levels.

The Vaughn Next Century Learning Center in Los Angeles offers one example of skill-based pay. Through a combination of self, peer, and administrator evaluation, teachers are rated on their skills in literacy instruction, language development, technology, inclusion, classroom management, and lesson planning, and can earn bonuses of up to $3,500 for demonstrated competence. Expertise in particular subject areas brings additional bonuses, and teachers can also earn a $4,000 award for national board certification (Chan 1999).

Odden and Kelley provide several models for structuring skill-based compensation:

1. Skill-based pay could be provided as increments to the regular salary schedule. That is, the bulk of the salary would still be determined by a combination of experience and education, but extra stipends could be provided for National Board of Professional Teaching Standards (NBPTS) certification or a second endorsement area.

2. A more radical model would replace automatic annual increments with increments based on a review of performance. In addition, the education-based lanes (based on degrees attained) would be replaced by demonstrated competencies, and there would also be a 5 to 10 percent annual premium for those who have attained NBPTS certification.

3. The third model takes a further step by replacing the existing schedule with an entirely skill-based system. Increases would be based on demonstration of specific skills; annual automatic increments would begin only after the teacher earned NBPTS certification (as an incentive to keep outstanding teachers in the profession). Teachers could also earn extra pay for demonstrating skill in a second content area.

4. The final model uses only competencies to determine compensation. Like the previous model, it bases pay on demonstration of competency in one or more content areas, and also adds the possibility of compensation for specific skills needed at the site level.

Although the concept is logical and appealing to many, competency-based compensation raises many questions at the implementation level.

1. How should different skills be valued? Some are presumably worth more, or are harder to achieve, than others.

2. Will the use of skill-based compensation add significant money to teacher earnings or will it simply shift the money around? Given the widespread recognition that teacher salaries are not competitive with other fields requiring a college degree, teachers may not be enthused about moving to a system that requires considerably more work to merely stay at the existing income level.

3. How will competency be measured? The easiest answer—and the least satisfying for an accountability framework—is to assume that attendance at workshops or attainment of certificates reflects competence, but as in the current system, that assumption is questionable. Alternatively, skills could be assessed in context (much as the NBPTS standards are). This would provide greater assurance of skill, but at a considerable cost, both in time and money.

4. Finally, does a skill-based system actually satisfy current definitions of accountability? Although use of specific skills is more defensible than college credits, such a system still relies on inputs (what teachers bring to the classroom) rather than outcomes (what students leave with). On the other hand, the growing recognition that motivation is not enough—that teachers and administrators must also have the capacity to change—suggests that a competency-based system can at least contribute to accountability efforts.

### Pay for Performance

What most people mean by "accountability" is results, specifically student achievement. Why not compensate teachers according to how their students perform? Why, the critics ask, can't schools be like the rest of society?

Despite the rhetoric, pay for performance is not that deeply ingrained in American practice. Aside from sales people on commission, factory workers on piecework, and small business owners, few workers live on a direct connection between results and pay. Teachers seem to dislike policies that put them in competition with peers, or that elevate one teacher over others, and unions are deeply suspicious, if not resistant, to performance-based proposals.

Clotfelter and Ladd (1996) claim that much of the resistance comes from the competitive nature of proposed programs, and they argue that any incentive system should measure achievement by schools rather than by individual classrooms. Odden and Kelley endorse this approach, saying that awards for group-based performance not only alleviate divisiveness but provide a powerful symbolic recognition of the collective effort required to boost student learning.

However, even group-based awards must be designed carefully if they are to have a motivational effect. Compensation experts offer the following advice to policymakers:

1. ***Define the desired result.*** Odden and Kelley note that organizations get more of what they measure, so districts should be sure they pay for the results they value the most. This will generally be some form of student achievement, es-

pecially in the core academic areas, but may also include increased attendance, lowered dropout rates, or greater parental involvement.

2. ***Choose the measure.*** Not all assessments measure the same things. A standardized test that measures factual knowledge would not be appropriate if critical thinking were the most desired goal. On the other hand, newer kinds of assessment that might better match the goals are still relatively unproven, making them questionable choices for allocating rewards (David Cohen 1997). Odden and Kelley recommend against the use of norm-referenced tests, which imply that not all students can achieve high standards. Finally, practical issues such as feasibility and cost have to be considered.

3. ***Set the target.*** What would be the criteria of success? How good must the results be to trigger rewards? While conceding they have no clear answer, Odden and Kelley say the ideal target will be above what is normally expected but not so far above that it seems out of reach. They also urge the use of "value-added" methods that measure gain in achievement rather than absolute scores that favor schools with advantaged students.

Even the relatively technical question of how to define *student achievement* has significant consequences. Clotfelter and Ladd, examining data from South Carolina, identified a number of possible scores that could be used to represent student achievement:

- the average absolute score of fifth-graders in the school
- average change in the test scores from one year to the next
- difference between the average fourth-grade scores and average fifth-grade scores in the school
- percentage change in scores
- "school gain index" (difference between the actual fifth-grade score and the "predicted score" based on previous-year scores)
- a variety of adjustments taking into account the socioeconomic background of students

Some of these outcomes are correlated with one another, whereas others are not. Thus, a school that does well by one measure may not fare as well on another. Yet a case could be made for using any of these outcomes.

Clotfelter and Ladd raise another issue: Will adjustments be made for socioeconomic factors? On the surface, the idea makes sense; virtually everyone acknowledges that schools with high numbers of at-risk students will generally emerge with lower achievement rates than will a homogeneous suburban school. Simply comparing absolute scores would be unfair to teachers working with the at-risk students. Although schools can use a number of strategies to compensate, each has its own disadvantages and complications.

One obvious idea is to measure growth during the year rather than the absolute level of achievement. However, this still doesn't negate the socioeconomic disadvantages some schools work under, since the factors that hold down achievement over a number of years also operate during each individual year. Regardless of quality of teaching, children from chaotic home environments are likely to learn less in any given year.

Another strategy is to compare actual scores with a "predicted" score. In essence, this means comparing scores with scores from similar schools. However, this may result in a subtle message that certain categories of children are not expected to achieve at high levels.

Cohen warns that policymakers need to make tradeoffs between fairness and clarity. A truly fair system would be impossibly complex: "A performance reward scheme of that sort could become the educational equivalent of Ptolemaic astronomy, with adjustments loaded onto adjustments until few could comprehend." Clotfelter and Ladd note that the formula used in South Carolina is as follows:

$$R_{it} = a + bR_{i,t-1} + c(R_{i,t-1}M_{i,t-1}) + dR2I^{,t-1} + eM2I^{,t-1} + u_{it}$$

While such formulas make performance rewards statistically sounder and fairer, they also divorce the process from everyday understanding, creating suspicion or just plain bewilderment. After describing an even more involved process used by the Dallas, Texas, schools, Clotfelter and Ladd note, "School officials neither understand the process nor have any idea what sorts of gains would have been required for them to achieve a high ranking."

4. ***Determine the amount of the award.*** How much money does it take to get the attention of teachers? The answer is not yet clear. Odden and Kelley suggest that $1,000-2,000 may well be sufficient, especially when combined with decentralized management that gives teachers more freedom of action. Given the skepticism of many teachers that awards will ever be paid (Kelley and colleagues), the promised amount may be less motivational than the actual delivery of the check.

Odden and Kelley recommend that awards be distributed as annual bonuses, not as additions to base salary. Increasing the salary will benefit the teacher for years to come, not just for the year the award was earned. Similarly, they favor uniform lump-sum payments rather than some percentage of salary.

5. ***Find ways to discourage "gaming."*** Performance-based systems emphasize results; in theory, how those results are achieved is not important. In the real world of schools, however, the means are as important as the ends. For example, narrowly teaching to the test may boost scores in the short term but undermine achievement in the long run. Similarly, educators being paid for high test scores may choose to exclude disadvantaged or at-risk students from the tests, or at least not include their scores in the average. A subtle version of this tactic is to discourage low-performing students from attending on test days, or simply not testing absentees when they return to school. The rule for policymakers designing performance-based systems is "beware the law of unintended consequences."

Do performance incentives for teachers boost student achievement? So far the question is unanswerable. Student performance is influenced by many factors besides teacher motivation, and accountability programs tend to change many of the variables at the same time. A typical initiative may highlight desired skills through new standards, add instructional time, provide better-focused professional-development activities, and provide students as well as teachers with incentives. If achievement goes up, it's almost impossible to determine how much of the impact is due to each of these changes.

We can, however, draw some conclusions about how well incentives influence teacher *motivation*. The Center for Policy Research in Education, after surveying schools using some form of pay for performance, concluded that teachers understood the accountability goals and were committed to them more than to other types of reform efforts (Carolyn Kelley and colleagues 2000). In addition, school performance was related to how well teachers understood the goals, as well as to capacity-building elements such as such as strong leadership and an effective information system that provided timely feedback on student progress.

However, the study also found that performance-based programs were less effective if the desired outcomes conflicted with other organizational goals (this was the case, for example, in magnet schools whose strong philosophical commitment conflicted with the performance goals). Programs were also undermined when the bonus was considered too small ($400-600) or when teachers were skeptical that the money would actually be paid. Finally, some teachers (especially those working with externally imposed goals or with continuous-improvement processes) reported higher stress with performance-based programs.

In general, Kelley and colleagues concluded that well-designed performance-pay programs work by focusing attention on desired goals as well as by providing concrete incentives. They recommend the following design elements:

- supportive district and principal leadership
- meaningful incentives (they suggest $2,000 a year)
- capacity-building programs
- goals that are set at a realistic, achievable level
- involvement of all parties in design of the criteria, so they are perceived as fair

## Incentives for Schools

Although teaching has often been criticized as an "isolated" profession in which practitioners operate as individualists, every teacher operates in an organizational context that shapes and influences the possibility of success. For that reason, some reformers believe that schools, not individuals, should be the target of incentive systems.

Accordingly, some states have established award programs in which the money goes to the school, to be spent on educational improvements as the staff collectively determines. For example, Kentucky sets goals for schools in terms of increased numbers of students scoring at the "proficient" level. Schools that exceed their goals by a certain level are eligible for awards, with the certified staff determining how the money is to be allocated (Tom Willis and colleagues 1999).

Policymakers in general, however, have been less interested in offering rewards for schools that succeed than in imposing school-based *sanctions* on those that fail. States are responding to low-performing institutions by limiting their freedom of action, taking them over, or even closing them. As of summer 2000, twenty-three states had established legal provisions for drastic state intervention in poorly performing school districts. In some cases, intervention is triggered by inept management, fiscal irregularities, or legal malfeasance, but at least fifteen states allow intervention for academic bankruptcy (Education Commission of the States 2000).

Intervention can take two forms: *takeovers* and *reconstitution*. In takeovers, the existing management team is replaced by state officials or state-appointed officials, but other employees are left in place, at least initially. Typically, states invoke this step only after the school or district has been given warning, with a probationary period in which they can rectify the problem. The state may require the school to develop an improvement plan, with specific targets; often the state will provide guidance for the school. Some states go beyond a takeover to wholly dissolve a district or school, deny accreditation, or put the school into the marketplace. Florida will offer vouchers to parents of students in failing schools.

In reconstitution, a low-performing school is essentially broken up and put back together with different pieces. School-site leadership is changed, and the teaching staff is either removed en masse or forced to reapply for their positions.

Does drastic intervention work? The evidence is mixed—and mostly anecdotal. In a report on state takeovers, the Education Commission of the States said, "For the most part, they seem to be yielding more gains in central office activities than in classroom instructional practices." Although new management can eliminate corruption and improve efficiency, it is much harder to raise student achievement. ECS concluded that takeovers "have yet to produce dramatic and consistent increases in student performance."

Similarly, a 1998 U.S. Department of Education report concluded, "To date, there are no conclusive data demonstrating that the threat of reconstitution is an effective motivator for change." The report noted that low-performing schools often have a long legacy of failure that cannot be overcome simply by replacing the teachers. The low expectations, deteriorating plant, and general demoralization are accepted by parents and community as well.

The American Federation of Teachers (1998) has supported reconstitution as part of a comprehensive, goal-oriented approach to reform. They argue that intervention works best when it:

- is based on high academic standards
- enforces high standards of behavior
- uses understandable and objective criteria for identifying low-performing schools
- addresses the needs of the school
- is based on research
- involves staff and provides them with the time and resources needed to make improvements

In no case should simply replacing staff be accepted as the solution.

The AFT cites the case of Corpus Christi, Texas, where the district and the union have worked together to disestablish—

and redesign—eleven schools. Low-performing schools were identified by a locally developed achievement test. Staff members at these schools had to reapply for their jobs; about a third of the original staff was rehired (the others were reassigned elsewhere in the system). The disestablished schools were designated as "special emphasis" schools and given extra resources that added instructional materials, created parent resource centers, and lowered the teacher-student ratio. The district created additional paraprofessional positions and added art and music specialists who were unavailable at other schools in the district. Teachers who agreed to work in these schools received extra stipends and were also given ten to fifteen days of professional development.

The results, after fifteen years, were that no Corpus Christi schools were considered low-performing, and a greater proportion of district schools reached the top two levels of Texas' accountability ratings than have schools in any other urban system in the state. Nine of the eleven disestablished schools were performing at or above the statewide average rate of improvement.

The Corpus Christi model raises an interesting irony in the accountability system. Reconstitution is clearly designed as a stick, something that teachers and administrators will strive to avoid. But here the schools were rewarded with more resources as a result of their low performance. Part of the explanation is that the Corpus Christi approach addressed the capacity issue rather than the motivational issue.

The U.S. Department of Education (1998) has provided advice for districts contemplating (or facing) reconstitution:

1. Strong leadership at the site is critical. Although successful schools can often coast for a while on the strength of past successes and faculty expertise, reconstituted schools are much more fragile. Principals of reconstituted schools should bring with them considerable experience and expertise.

2. Successful transformation of low-performing schools requires a clear break with the past—not only with new staff, but with new structures (such as creating a specialty school).

3. The rebuilding effort must incorporate high standards for student achievement.

4. Professional development and capacity building are crucial. Not only must the school implement new patterns of instruction, but teachers in reconstituted schools tend to be younger than in most schools and can benefit from more plentiful offerings of professional development.

5. Watch for unintended consequences. For example, moving staff around affects not only the reconstituted school, but other schools as well.

6. State and district leadership is essential.

## Motivation and Incentives

As most of the public sees the matter, incentives are the heart of the new accountability system; almost no one questions the assumption that the right kind of incentives will energize students and teachers to work harder and produce better results. This assumption, based largely on behaviorist principles, seems unchallengeable at the common-sense level. People *do* work harder and faster when they have a chance to earn a desirable reward; positive reinforcement *will* grab attention and focus energy.

However, those generalizations can be undercut by the complexities of human behavior in real-world settings. In particular, three issues can blunt the motivational power of performance incentives.

1. *Not all incentives are equally rewarding.* Teachers, by and large, are not driven by money. Like most people, they welcome it, and in the collective-bargaining arena they will work hard to get more of it. But they did not enter teaching with visions of wealth, and money does not ignite their passion and mobilize their energies. Instead, they live for the daily small victories of student growth, measured not so much by test scores as by a gleam in the eye that says, "Now I get it!" Programs and strategies that increase such moments will be more powerful motivators than a bonus for raising test scores.

Significant monetary incentives do improve motivation by signaling that policymakers are serious about reform, making teachers more willing to invest time in the new approaches. And bonuses add a satisfying punctuation mark to teachers' efforts to help their students reach new standards. In their review of North Carolina's performance-compensation system, Henry Johnson and colleagues (1999) concluded:

> Although the bonus itself did not appear to be perceived as an incentive in and of itself, as part of an accountability program with rewards, sanctions, and assistance it appeared to have some positive value and to assist in focusing the goals for a school.

But by themselves, performance awards are unlikely to turn teachers into enthusiastic supporters of standards or move them to adopt dramatically new approaches in the classroom.

2. *Before responding to incentives, many people perform a simple cost-benefit analysis.* The value of an incentive goes down as the effort required to earn it goes up. Implementing standards-based instruction demands substantial time and energy, as well as the willingness to change deeply engrained teaching habits. Were money the only consideration, a $1,000 bonus would entice few teachers into embracing standards.

3. *No matter how attractive the incentive, motivation will be limited by perceived capacity.* That is, if teachers believe that their students are incapable of achieving the new expectations, or if they believe they themselves lack the necessary resources, administrative support, or instructional strategies to get the job done, they will have little reason to invest extra energy in standards.

In short, financial incentives appear unlikely to transform classrooms. As part of a comprehensive accountability system, they can focus attention and add credibility to reform efforts. But schools are not stock markets, and teachers do not judge their success by the number of dollars they can squeeze from the system.

## Conclusion

Despite the political and philosophical appeal of incentives, states and districts have moved slowly to implement rewards and punishments. The technical questions are formidable, and the political implications are always lurking in the background.

(As many policymakers have discovered, consequences have consequences.)

Yet the push for accountability has nudged school systems into experiments that would have been impossible just a decade or two ago. A recently arrived time traveler from the 1970s would be astonished to see teacher unions willing even to discuss performance-based pay, much less negotiate agreements that allowed it. Whether this experimental period will evolve into a permanent restructuring is impossible to predict. The answer depends on the viability of standards-based accountability, on the solutions to some daunting technical questions, and on the numbers of jobs lost and diplomas denied.

More important, the evidence thus far suggests that incentives alone are not the complete answer. Used properly, they can call attention to what is important, modify teacher and student behavior, and create enthusiasm for change, but they do not magically endow participants with the tools needed for meaningful change.

Chapter 7 examines the remaining piece of the accountability puzzle: what schools can do to build capacity for academic excellence.

# 7

# Building Capacity for Improvement

*High standards will avail little unless teachers and administrators develop the capacity to do more than simply redouble their efforts with the same old methods.*

Today's accountability systems are built around the assumption that low-achieving schools suffer from unclear goals and low motivation. In theory, well-designed standards will sharpen the goals, and the combination of assessment, incentives, and public reporting of results will boost motivation.

While early experience with accountability has provided examples of the process working in just this way, it is also clear that some schools do not immediately improve. Why?

Perhaps the laggard schools simply need additional time. We don't expect all students to progress at the same rate, so why should we expect teachers and administrators to do the same? Or perhaps, in a more pessimistic interpretation, some schools are beyond hope, populated by burnt-out teachers and hidebound principals who believe they can deflect, subvert, or simply wait out this latest reform fad. If so, then the best strategy is to follow the logic of accountability by dismissing the underperformers and reconstituting the hopeless schools.

However, there is at least one other possibility: that failing schools are staffed by caring, competent professionals who lack the opportunity to engage fully with the meaning of new standards and to develop the strategies needed to meet the challenge. The new standards are ambitious and rigorous, pushing teachers into uncharted territory.

In the words of Linda Darling-Hammond and Milbrey McLaughlin (1996), the reform agenda requires most teachers "to reconceptualize their practice, to construct new classroom roles and expectations about student outcomes, and to teach in ways they have never taught before and probably have never experienced." More specifically, Ann Lieberman has listed the tasks as building new roles, inventing new relationships, creating new structures, working on new tasks, and creating a culture of inquiry.

With that kind of agenda, success will depend on the ability of teachers and administrators to learn what experience has not taught them—and perhaps to unlearn some things that it has. Clearly, as the Southern Regional Education Board (2000) has emphasized, fulfilling the promise of the standards movement will require well-focused professional development. However, it may be a kind of professional development radically different from previous models.

This chapter examines the learning demands that standards-based accountability puts on teachers and administrators and discusses ways that practitioners can increase the system's capacity for meeting new standards.

## Supporting Teacher Learning

No matter how highly motivated, people cannot master a new task without the relevant knowledge and skills. Increasingly, accountability advocates are recognizing that setting high standards will be futile unless teachers and administrators develop the capacity to do more than simply redouble their efforts with the same old methods.

Jennifer O'Day and colleagues (1995) note that teacher capacity has four dimensions:

1. *Knowledge*. Teachers need a firm grasp of their subject matter as well as strategies that convey the essence of that subject.

2. *Skills.* New standards require new instructional strategies, many of which are unproven, unfamiliar, and uncomfortable.

3. *Dispositions.* Teacher attitudes are the key to teacher actions. If teachers believe, for example, that students are not capable of meeting new standards, then they are unlikely to have much interest in new methods. Or, if they distrust change, even well-researched strategies will hold little interest.

4. *Views of self.* How teachers define their craft influences their actions. If they see themselves as managers, they may fret that new strategies will not fit into the daily routine. If they see themselves as personal counselors, they may worry that higher standards will create new barriers for disadvantaged students. If they are focused on academic achievement, they may either welcome the rigor of the new standards or disagree with their content.

Thus, increasing teachers' capacity requires that they engage in multiple types of learning, often simultaneously. Professional development, especially for standards-based reform, will never be a matter of simple skill transfer.

## The Demands of Reform

The public typically views standards-based reform as a matter of doing better what schools have tried to do all along. Since some schools are demonstrably successful, we simply need to make sure their model is replicated throughout the system. The apparatus of accountability is just an elaborate way of making sure that this happens.

Contrary to this common perception, the nature of new standards may actually require all teachers to critically examine their existing practices, reflect deeply, and possibly unlearn some of what they already know. For example, Deborah Ball (1994) has documented the complexities of teaching math according to new standards that emphasize reflection and respect for students' ways of thinking. She notes that under the best of circumstances it is often difficult to evaluate student thinking and respond appropriately. One of her third-graders argued that 5/5 was more than 4/4, because if you divide a cookie into five pieces you can give a piece to five friends, but with four pieces one friend will come up short. Pondering this episode, Ball says:

> Sheena was being creative. And some aspects of her answer were "right." But her nonstandard approach had actually changed the question. And her response to the original question was "wrong." What should be the "right" answer for me here? To this day, that remains uncertain. The slogans "teaching for understanding" and "mathematics for all" are a lot more complex when viewed close up.

Confronting these uncertainties, teachers find the answers elusive, not just because the issues are difficult but because their experience has provided them with few helpful models (Ball and Cohen).

People generally learn to teach in three broad phases. First, even before they know they want to be teachers, they spend years in what Dan Lortie has called "the apprenticeship of observation," quietly (and quite unconsciously) absorbing lessons about life in classrooms and about what it means to teach.

Entering formal training (the second phase), teachers are sometimes exposed to state-of-the-art theories and encouraged to step back and think reflectively about teaching, but ideas that look good in the peaceful precincts of a college classroom are often overwhelmed in the pressure cooker of real classrooms. In the end, teachers learn most of their craft on the job, usually by following the model of their peers.

Thus, the craft of teaching is governed not so much by formal theories and systematic strategies as by "folkways"—assumptions, practices, and traditions that seem like common sense to practitioners (Margaret Buchmann 1987). Among the most persistent folkways are recitations, seatwork, and lectures. Although there are good reasons for the existence of these patterns (they would not persist if they weren't helpful in some way), they have a taken-for-granted quality that makes it harder to see other possibilities.

Ironically, professional-development activities often match the folkways by using a simple learning paradigm that makes this assumption: Teachers need additional knowledge and skills that can best be acquired from experts through a process of telling and showing (James Spillane 2000). In this paradigm, the professional-development curriculum typically consists of a wide variety of topics that are distantly related; a session on cooperative learning may be followed by an examination of math standards followed in turn by a demonstration of writing rubrics. This view sees teacher learning as cumulative, each session constituting one more brick in the edifice of knowledge. The coherence of these diverse topics depends on teachers' ability to integrate them into their practice.

The irony, of course, is that the old paradigm is often the vehicle used for urging teachers to develop new paradigms, a discrepancy they seldom miss. In the words of one exasperated workshop attendee, "Nothing like six hours of leaden lecture on the evils of lecture-style teaching to rub the old ganglia raw" (Bryan Jones).

Increasingly, critics are taking aim at the typical hit-and-run workshops that offer information and inspiration, but little depth or follow-through. This scattershot approach—popular with schools trying to keep up with the latest fads—may be "the professional equivalent of yo-yo dieting" (Deborah Ball and David Cohen 1999). Instead, drawing on newer theories of cognitive learning (Spillane), professional-development theorists argue that sustained reform requires a "new paradigm" that turns traditional professional development on its head (Ball and Cohen; Dennis Sparks and Stephanie Hirsch 1997; Willis Hawley and Linda Valli 1999). The new model has several key features:

1. *It focuses on organizational, not just individual, change.* The old assumption was that if you could change teachers, you could change the school. But even when teachers are ready to move in a new direction, the school itself may put up barriers to meaningful change.

2. *It uses a central vision and coordinated strategy rather than piecemeal efforts.* Schools have so many needs that they often adopt a smorgasbord approach that diffuses their efforts. While professional development is never mono-

lithic, a clear focus improves the chances of achieving meaningful change.

3. *It is embedded in everyday activities, not detached in after-school workshops or summer institutes.* Ball and Cohen note, "Teaching occurs in particulars—particular students interacting with particular teachers over particular ideas in particular circumstances." General strategies provide useful starting points, but must always be adapted to the needs of one's own classroom.

4. *It uses student learning as the ultimate measure of success.* When teachers leave a workshop enthused and excited, professional development is off to a promising start, but it crosses the finish line only when that enthusiasm is translated into improved academic performance.

5. *It views teachers as competent professionals rather than as empty vessels to be filled.* While the new paradigm emphasizes how much teachers have to learn, it also recognizes that they already know a lot. The goal is not to replicate the work of outside experts but to extend their own expertise.

6. *It emphasizes reflection rather than replication.* In part, this means that teachers should be exposed to research-based theory that provides perspective on their practice, but even more it means that teachers should have opportunities to think deeply about what happens in their own classrooms.

7. *It is collegial and collaborative.* When teachers work together, they develop a shared language, challenge one another's perceptions, and create mutual respect. Hawley and Valli say, "Without collaborative problem solving, individual change may be possible, but school change is not."

8. *It is continuous, a process rather than an event.* Meaningful change does not occur overnight; leaders who initiate teacher learning activities may not see tangible changes in practice for three to five years.

## Practice-Based Teacher Learning

In states that have begun assessing progress in implementing their new standards, the early results have often been disappointing. Teachers, struggling to help their students with the challenging expectations and seeing so many fail to do so, have sometimes argued that the expectations are unrealistic or developmentally inappropriate. For example, the Arizona Education Association and the Seattle Education Association have resisted high-stakes testing partly because of their belief that some of the standards are developmentally inappropriate (www.fairtest.org/union.union.html).

Respect for teachers' expertise requires us to take that possibility seriously. It may well turn out that in some cases reformers have aimed too high. Yet it is also possible that when teachers say students cannot meet the expectations, they actually mean that even their best strategies are not working. That is, the goal may be achievable, but the tools are lacking.

The most ambitious standards ask students to think at a sophisticated level, to "actively try to solve problems, resolve dissonances between the way they initially understand a phenomenon and new evidence that challenges that understanding, put collections of facts or observations together into patterns, make and test conjectures, and build lines of reasoning about why claims are or are not true" (Thompson and Zeuli 1999). Students must not just learn to think, they must "think to learn."

To achieve this result, teachers must respond in particular ways:

> It requires posing questions framed with one eye on students' existing ideas and the other on the ideas to be learned; listening with one ear trained in each of these directions; asking students to explain their thinking; asking other students whether they agree or disagree, and, in any event, why; deciding which unexpected turns of thought to pursue and which to ignore for the moment; and a great deal more. (Thompson and Zeuli)

Thompson and Zeuli argue that accountability systems are not sufficient to bring about the right kind of teacher learning. Standards and assessment may provide guidance on the desired outcomes along with feedback on how students are doing, but they do not provide a readable roadmap showing how to get there. Even when teachers enthusiastically embrace new standards, they may adapt rather than adopt, integrating bits and pieces of the new ideas into their existing practice. What results is teaching that has some of the surface features of the new ideas but is still dominated by habitual ways of thinking.

None of this suggests that teachers are dense, or resistant to change. If they habitually tinker with new ideas rather than adopting them en masse, it is because incremental change allows them to preserve the always-delicate order of their classroom. Teachers' daily agendas are filled with more than lesson plans; they must manage an extraordinary flow of events while meeting the burgeoning social and emotional needs of their students. Out-of-the-box alternatives may appeal to theorists and consultants, but teachers view them with a skeptical eye.

So how can professional development escape the pull of tradition and provoke teachers into rethinking their craft? Advocates of the new paradigm say the answer is to embed professional development in teachers' everyday practice, focusing on the students, curriculum, and classroom conditions teachers daily confront (Ball and Cohen; Sykes 1999; Thompson and Zeuli). Just as student learning occurs in the classroom, so must teacher learning.

Practice-based teacher learning involves several elements:

1. ***Development activities are chosen for their likely impact on student learning.*** For example, a strategy with a successful track record (whether demonstrated through research or local experience) is easier to justify than an unproven idea. Sykes says that those who put forward proposals for professional development should map out how the activity is likely to lead to improved student learning.

In a standards-based system, a focus on student learning implies a focus on standards, no matter how appealing other topics may be. One Washington State principal put it this way:

> So we had to talk a little bit about for instance some pet projects that people like to do because they're nice, warm, fuzzy things and they've always done them. We really had to ask the question, "Well, does it get us where we need to go? Is time better used maybe in some other project?" (Robin Lake and colleagues 2000)

2. ***Activities are centered on specific curriculum content.*** According to Sykes, some studies suggest that teachers' practices are most likely to be changed when development focuses on the particular content and goals their own students are struggling with. Thus, rather than a generic workshop on math manipulatives, teachers are more likely to benefit from activities that examine how their fourth-graders are currently thinking about fractions.

3. ***Teachers regularly work together to assess their students' work.*** The standards movement ultimately depends on someone's ability to judge whether students have met the expectations. When the standard calls for complex thinking, this evaluation will not always be easy. By collectively examining student work, teachers can deepen their understanding of what the standards mean and how close to the target their students are coming.

The Annenberg Institute for School Reform has developed structures and protocols for conducting this sort of evaluation. A typical protocol begins with a teacher selecting a piece of student work to share with a small group of colleagues, parents, and students; at this time the teacher explains the standard that best applies to the activity. The teacher then describes what the student was asked to do, after which the group examines the work closely, making observations and raising questions. Finally, the group revisits the standard and applies it to the work at hand. Has the student met the standard? In the process of answering that question, participants come to a better understanding of the standards, student thinking patterns, and their own teaching (Don Glass 2000).

Another example is suggested by Ball and Cohen. In their scenario, middle-school teachers examine math achievement to try to determine why scores on the state-mandated performance assessment are falling. The teachers begin by examining a sampling of student responses on the assessment, but find they don't completely understand what the students are being asked to do. So they work through some of the questions themselves, paying close attention to how they arrive at the answers. Having done this, they are in a better position to see how their students arrived at answers, and where their thinking differs. Eventually, as they gain an understanding of the major stumbling blocks their students encountered, they begin to reach outside the school to look for focused assistance in dealing with specific issues.

4. ***Teachers collectively examine and reflect on their own teaching.*** Teachers' daily life is filled with frenetic activity, leaving little time for reflection, with only blurry impressions of what happened and why. By documenting their practice (through videotape, for example) and sharing it with others, they can capture the complex interactions among teacher, students, and curriculum, and invite their colleagues to pose questions and offer critiques.

5. ***Professional development is data-driven.*** Standards-based accountability generates a wealth of data that should be used for improvement, even if it means revamping cherished activities. Well-designed development activities often unleash teacher creativity and excitement, but if the data fail to show improvement, another direction is needed. Bruce Joyce and Emily Calhoun (1995) describe an elementary school that developed a full range of innovative writing activities, such as visits from children's authors and a Write-Night Sleep-in. After teachers found that writing improved relatively little for all the effort that had been expended, they then decided they needed to focus on classroom instruction.

In a profession often characterized as "private" or "isolationist," subjecting one's work to public scrutiny takes a certain amount of courage, and it may take time for teachers to develop a level of comfort with such activities. But without grounding teacher learning in the everyday work of teachers, professional development becomes little more than "an exchange of buzzwords or slogans" (Ball and Cohen).

## Learning Communities

Some schools, taking their cues from recent private-sector experience with "learning organizations," have worked consciously to establish what they call "learning communities" or "communities of practice." In a learning community, "teachers in a school and its administrators continuously seek and share learning, and act on their learning" (Shirley Hord 1997). The goal is not just to collaborate, but to improve the group's effectiveness in helping students.

Hord says that effective learning communities have five key attributes:

1. *Supportive and shared leadership.* Learning communities reject the notion of the "omnicompetent" principal in which school leaders evaluate staff needs, determine appropriate professional-development activities, and implement activities. Instead, leaders view themselves as colearners with the rest of the staff, and teachers take responsibility for their own learning (although in a collegial rather than individualistic way). This kind of leadership requires principals who can share authority, facilitate the work of others, and participate without dominating.

2. *Collective creativity.* Community implies collaboration, which in turn helps participants see beyond their own intellectual frameworks and consider new forms of practice. When a group of fourth-grade teachers get together to consider their practices of classroom assessment, their reflective dialogue is likely to identify new forms of assessment that better measure state content standards. Penelope Wald and Michael Castleberry (2000) characterize this process as synergy, noting that participants are often surprised by what emerges from the group's deliberations.

3. *Shared values and vision.* Productive learning communities share a common vision centered on student learning, and they use this vision to guide their instructional and curricular decisions.

4. *Supportive conditions.* For learning communities to thrive, the physical and social environments must support teacher learning. Proximity to colleagues, schedules that allow collabo-

ration, resources to carry out training, and, above all, time for staff to reflect on their practice are essential. Just as important are the human qualities that make professional development thrive: openness to change, willingness to accept feedback, mutual respect and trust, and robust socialization processes.

5. *Shared personal practice.* Learning communities are intensely collaborative, with teachers not just getting together for occasional meetings on neutral ground, but observing and critiquing one another's classroom practice. In a field that often values privacy, teachers in learning communities are not afraid to observe others, open their own work to inspection, and objectively consider feedback.

Hord notes that research on learning communities provides compelling evidence that the results include reduced teacher isolation, higher morale, deeper understanding of the learning process, and, most important, improved student learning.

Taking a school from an individualistic culture to a community of practice is a formidable challenge, and researchers are still in the early stages of identifying the ways that leaders can make the transformation. Hord cites the work of Louis and Kruse (1995) as offering some clues. In the urban schools they studied, leaders of schools with strong professional community seemed to follow half a dozen principles:

1. *Physically and psychologically, principals led from the center.* That is, they were both visible and accessible to teachers, and, while leaving no doubt they were assertive leaders, they were willing to give up some of the traditional leader behaviors such as always running the meetings.

2. *Principals provided classroom support.* As teachers struggled with the complexities of student learning, administrators made sure they had assistance, either through collective dialogue or outside expertise.

3. *Principals held a vision of professional community.* The community in their school was no accident; they believed firmly in it and articulated it to others.

4. *Principals created a culture of high intellectual quality.* They continually brought in new ideas and people to stimulate teachers' thinking; they provided concrete data to help teach-

ers judge program results; and they promoted action research by teachers.

5. *Principals managed conflict.* They recognized that conflict in a learning community is not only inevitable, it is healthy, and they addressed disagreements openly, through dialogue and debate.

6. *Principals practiced inclusion.* They worked to pull in those at the periphery as well as the enthusiasts, giving all teachers a chance to feel they were part of the community.

## Learning Communities Beyond the School

When teachers are unable to find answers to their questions within their own school community, they sometimes reach out to find like-minded counterparts at other schools. The interaction may be as casual as a lunch-table conversation at a conference, but occasionally it ignites into something more extensive and more enduring. In the last two decades, teachers have increasingly been able to form networks centered around shared concerns and interest.

A network can be informal and spontaneous, as in a computer mail group, but it can also take on a physical bricks-and-mortar presence (a teacher center), offering a home away from home for teacher-learners. Because they are often established outside the normal chain of command, networks carry an aura of autonomy and professionalism. Networks engage in learning that their members want, not learning that has been imposed from the top. They provide a variety of activities that go far beyond the chalk-and-talk format of conventional inservice activities, create discourse communities that allow the development of a common vocabulary and professional exchange among like-minded teachers, and develop leadership (Ann Lieberman and Milbrey McLaughlin 1996).

More concretely, networks and centers offer a forum for discussing standards and assessments, providing assistance for specific problems, developing curriculum, promoting National Board certification, and conducting action research (National Foundation for the Improvement of Education).

Networks are not without problems, however. An inherent paradox is that networks may draw teachers away from the communities of practice in their own schools. In addition, because they often operate without permanent or institutionalized funding, their stability is always at risk. And ironically, network success may lead to growth and institutionalization, sometimes resulting in structure and hierarchy that negate the autonomy that make them successful (Lieberman and McLaughlin).

## Structuring the School for Teacher Learning

A school's capacity for learning lies not just in the talents and dispositions of its teachers, but in its organizational structure and climate (O'Day and colleagues). Teacher learning does not occur in a vacuum. Highly motivated teachers can be driven to despair by an unresponsive organization, while indifferent teachers can renew their enthusiasm in a school that values and supports their learning.

Judith Warren Little (1999) notes the lack of comparative and longitudinal research on how workplace features affect teacher learning. But based on studies of innovations and current best thinking on organizational change, she suggests several basic strategies to make schools learning-friendly for teachers.

First, echoing the views of many others, she says that cultivating collective efforts to evaluate student work offers a rich potential for enhancing teacher learning and promoting inquiry. While the means of sustaining such work are not yet clear from the research literature, structural changes in the workday can help. For example, providing teachers with common planning time at least allows the opportunity for interaction.

Second, organizational issues such as teacher assignment and time usage have a major impact on teacher learning. When teachers are placed in a subject or grade level that is a poor match for their skills, their lack of confidence is likely to result in conservative and inflexible teaching. (Unfortunately,

Little observes, schools routinely place the most inexperienced teachers in the most difficult assignments.)

More positively, schools can use assignment practices to stretch teachers' capacity for growth. Little cites a high school English department that expected each teacher to be able to teach any of its courses. Since no one "owned" courses, the entire department developed a sense of collective responsibility for the curriculum. As a bonus, beginning teachers had access to an ample supply of advice and materials.

Third, time and money for professional development are always in short supply, so leaders need to ensure that the resources are allocated wisely. Little describes a wide range of possible activities that go beyond the typical one-shot workshops: planning days, teacher research teams, student-assessment activities, staff retreats, classroom and school visitations, computer technology, subsidized participation in summer institutes, and involvement in regional networks.

Fourth, providing teachers with valid and timely feedback on results will stimulate learning. Evaluations of teacher performance, which are perfunctory in many schools, can become productive learning experiences when enriched with portfolios, peer review, and self-assessment. Assessments of student learning can also promote teacher learning as long as the staff has structured opportunities for collaborative analysis and reflection.

Finally, Little argues that structural conditions alone will not be sufficient unless they are backed up by the school's values, beliefs, and norms. Even when the structure is supportive, people are more likely to take their cues from casual hallway conversations and the multitude of small daily actions that make up school life. For example, a well-designed teacher-evaluation event will have little impact if accompanied by a "file and forget" mentality.

## Finding the Time

Universally, practitioners and researchers say that the greatest roadblock to professional development is lack of time: time for meeting, time for reflecting, time for developing new instructional strategies. The teacher's day is not only jam-packed with activities and responsibilities, it is also fragmented into discrete blocks less than an hour long. If even two teachers want to meet, the odds are against their having a common time; if a committee of six needs to meet, their only hope is to squeeze out some time at the tail end of the day, when energy is at low ebb.

Mary Anne Raywid (1993) has identified three basic strategies for freeing up time:

1. *Add more time to the school day or year.* This is the most obvious, but most expensive, solution. For example, some districts have simply built three to five development days into the annual contract.

2. *Reallocate existing time.* Many schools have traded instructional time for development time; for example, school may be dismissed at noon once a month to allow teachers a full afternoon of development time. Although relatively easy to implement, such programs may butt up against state requirements for minimum instructional time. In addition, they are not popular with parents who have to alter child-care arrangements on those days. And other citizens may view this as "time off."

3. *Alter staffing utilization patterns.* Even if time remains the same, it can be used more productively by changing schedules or instructional arrangements. Raywid cites a number of examples:

- One school used a team of subject specialists (art, music, physical education, and so forth) to present half-day integrated lessons to classes while the regular teachers were working on professional development activities.
- A New York school gave teachers working on a new curriculum a common lunch hour followed by preparation period, giving them a total of ninety minutes of shared time each day.
- Some larger districts have increased class sizes by one or two students, allowing them to support teams of permanent substitutes who can cover classes while teachers are engaged in professional development.

Even if the time can be found, it still has to be used effectively. Raywid recommends that schools use the following principles in structuring their development time:

1. Professional development should be prime time, not something tacked on at the end of a busy day. Anyone who has sat in a classroom all day and watched teacher and students slowly wear down into a state of tired crankiness will understand why the end-of-day approach is often ineffective.

2. Development time must be sustained. Given the kind of reflection that is required by standards-based reform, a single class period is barely enough time to properly formulate the questions, much less seek out answers.

3. Even when compensated, teachers should not be asked to take all their development out of weekends or vacation time. For one thing, time away from the job is good therapy for weary professionals. Moreover, schools that rely on vacation time for development activities send an unmistakable message: This activity is not important enough to be done on school time.

4. Professional development should not rob students of good instruction. Teachers themselves are often reluctant to leave their classrooms in the hands of others, especially if there has been little planning for what will go on in their absence. Simply turning the class over to a substitute will not be popular.

## Governing for Learning

As teachers have all too frequently discovered, changes in school structure can be superficial, masking the fact that business as usual continues behind the scenes. Adding resources, altering schedules, and promoting collaboration will provide the platform for improved teacher learning, but these changes must also be accompanied by deep changes in attitudes about governance, power, and leadership.

Researchers are just beginning to study the relationship—apparently complex—between leadership decisions and teacher learning. On the one hand, Helen Marks and Karen Seashore Louis (1999) conducted an extensive study that found strong correlations between teacher empowerment and teacher learning. Empowerment was measured by the degree of shared decision-making and influence over such things as discipline, curriculum, textbook selection, and school policy. When teachers believe themselves to have considerable professional autonomy, they are more likely to be involved in ongoing collaborative learning and innovation.

On the other hand, Andy Hargreaves (1994) notes that good-faith efforts to facilitate collaboration can fall flat when not enough attention is paid to teacher needs. "Contrived collegiality" is what happens when administrators try to encourage teacher learning—but on the administrators' terms. For example, principals may arrange for teachers on an instructional team to have a common preparation period, only to find they use the time for other purposes. Hargreaves says the teachers are not resisting the idea of working together; rather, the officially designated time does not lend itself to effective collaboration (for example, the preparation time is too short for productive work). Hargreaves says that leaders will get better results by paying attention to the specific working conditions and needs of their teachers and by focusing on facilitating collaboration rather than trying to control it.

Hargreaves has also pointed out that collaboration can divide. When small groups of teachers habitually work together—and only with one another—schools become "balkanized," with multiple small groups developing their own philosophy and ethos and reinforcing one another's views rather than challenging them. This pattern is especially common at the secondary level, where subject specialization and departmentalization continually pull small groups of teachers into similar orbits. Hargreaves says that leaders are likely to have little luck eliminating this pattern, though it can be supplemented by structuring more inclusive groups. For example, schoolwide committees could require representation from each department.

Finally, Linda McNeil has pointed out the tension between accountability as bureaucratic mandate and accountability as educational reform. McNeil notes that some advocates of accountability are motivated by a "cost-accounting" mentality aimed at tightening controls and making sure the public is getting its money's worth. This form of accountability emphasizes efficiency rather than effectiveness, and focuses on standardized, easily measured behaviors. The result, according to McNeil, is a "dumbed-down," test-driven curriculum that squeezes out the best kinds of teaching.

In the Texas magnet schools McNeil studied, teachers had created rich, meaningful programs of study, only to be forced into impoverished instruction based on administration-mandated "test prep" materials. Invariably, the materials were superficial and fragmented, focusing only on basic skills, with no indication that the content was important enough for students to remember. In some cases, teachers were required to use the materials from September through March, when the statewide tests were given.

Many teachers responded with "double-entry" lessons that presented the required material in a perfunctory way but quickly moved on to the more meaningful material. Others created elective courses that were not yet covered by the standards. But inevitably, many teachers were forced to delete some of their best lessons to make room for the mandated materials; they recognized the negative consequences for their students who failed the test, and had to balance that against their conviction that the test-related material was essentially worthless. In addition, scarce resources were diverted to commercial test-prep materials. For both students and teachers, the system alienated them from their own sense of best practices.

In a strong mandate-driven system, the dilemma faced by these teachers was unavoidable, but it was sometimes worsened by principals or district administrators establishing their own mandates, and saying, in effect, "I don't want to hear about what you're doing unless it's about the test." Clearly, a key leadership challenge is to act as a buffer between the faculty's sense of professionalism and the more rigid aspects of the accountability system.

When policymakers first put schools on the path to standards-based reform, they saw it as an implementation problem, not a learning problem. With clear standards and meaningful incentives, schools would find a way to get the desired results. But as practitioners began responding to the challenge, they quickly discovered that questions were far more plentiful than answers, and that they would have to expand their toolkits to get the job done. Field-tested models were (and remain) scarce, and they cannot always be transported from one site to another. For now at least, schools will have to learn their way to success.

## Professional Development and Motivation

Professional development has remained in the shadow of standards, assessment, and incentives, promoted mostly by educators who recognize the complexity of reform. Policymakers tend to see it as an auxiliary function, a simple matter of skills-transfer from those who have solved the standards problem to those who are still struggling. In their view, it is not a cause, but a consequence, of motivation: Once the standards, assessments, and incentives have energized teachers, they will turn to professional development for answers.

However, the works cited in this chapter suggest that teacher learning can itself be a source of motivation, both because of the capacity it adds and because of the intrinsic satisfaction it provides.

As noted earlier, perceived lack of capacity undermines motivation. No matter how attractive the incentives, humans are generally disinclined to work toward goals they believe to be out of reach. Some teachers—a minority so far—are openly skeptical that their students can achieve the new standards. Most, perhaps encouraged by early gains on test scores, seem to have adopted a wait-and-see attitude. If the initial progress stalls out after a year or two, skeptics may form the majority. At that point, only intensive teacher learning that is centered on the standards can develop capacity and rekindle hope.

The new forms of professional development focus not on the agenda of outside experts, but on the concrete instructional issues that teachers face each day. When they can concentrate on the problems that engage their best thinking, work collaboratively with colleagues, and observe the results in their own classrooms, teachers are much more likely to pursue standards-based reform with a sense of satisfaction and even enjoyment.

# 8

# Conclusion: Unanswered Questions and Leadership Imperatives

*The fact policymakers have managed to get fifty state bureaucracies headed in approximately the same direction at the same time is an astonishing achievement.... Should educators and parents start to believe that policymakers will not follow through, the steam may go out of the accountability movement.*

Almost twenty years after *A Nation at Risk* started it all, even the most cynical onlooker has to admit that the accountability movement has staying power. Since 1983, policymakers have steadily pursued the idea that schools must show better results, and school boards, administrators, and teachers have slowly come to realize that it is not just a passing fad.

But if educators now accept accountability as a reality, they still view it warily, as something akin to an unwelcome relative who has arrived, uninvited, for a long stay, leaving the host torn between the obligations of courtesy and the desire to get back to a normal life,

Most schools have made a good-faith effort to respond to the expectations, revising curriculum, reallocating time, and searching for new instructional strategies. For some schools, the efforts have paid off with improved assessment results. But even in successful schools, teachers and principals remain ambivalent, wondering how much of a rise in test scores would justify the time, money, and energy being poured into accountability efforts.

Thus far, research does not tell us much about the effectiveness of standards-based accountability. Few states have fully implemented their systems, and it will be years before trustworthy evidence is available. At best, the current literature allows us to draw a few cautious conclusions.

We now know that we can articulate rigorous standards and design assessments that measure those standards. We know that parents and the public will support measuring student performance against those standards (albeit with a few reservations about standardized testing). We know that educators are responding to the combination of standards, assessment, and public reporting by actively seeking ways to meet the new expectations. And we know that some schools have shown improvement.

Those conclusions, modest as they are, suggest that standards-based accountability has a future. The fact policymakers have managed to get fifty state bureaucracies headed in approximately the same direction at the same time is an astonishing achievement, and it attests to the depth of the public's desire to make the system work.

## Bush Administration Proposal

This rough consensus across states was reinforced in the first week of the George W. Bush Administration with the release of the president's education proposal, *Leave No Child Behind,* which relied heavily on the assumptions and recommendations of accountability advocates (see table 8.1). While not imposing a whole new accountability system on states, the Bush platform added federal resources—and teeth—to existing state efforts.

The plan did not directly create new standards, but did require that states adopt standards in reading, math, science,

**Table 8.1**

## Bush Administration Proposals Corresponding to Recommendations of Accountability Advocates

| Accountability Components | Bush Education Proposals |
|---|---|
| I. Clear, rigorous standards | • States must establish standards in reading, math, science, and social studies. |
| II. Assessments aligned to standards | • States must require annual assessments of every child in grades 3-8.<br>• State results will be confirmed by sampling of fourth- and eighth-grade students on the NAEP.<br>• States may design their own assessments, as long as results would be comparable from year to year. |
| III. Public reporting of results | • States are required to report assessment results to parents.<br>• States are required to report to public results disaggregated by race, gender, English-language proficiency, learning disability, and socioeconomic status. |
| IV. Rewards and sanctions | • States or districts will determine which schools have not made adequate yearly progress.<br>• If a school shows inadequate progress after two years, district must take corrective action and must allow students in the failing school to transfer to other public schools.<br>• If a school shows inadequate progress after three years, disadvantaged students may use Title I funds to transfer to higher performing public or private school or receive supplemental educational services from a provider of choice.<br>• Schools that narrow the achievement gap will be given bonus funds.<br>• States that fail to adequately narrow the achievement gap on math and reading tests will lose a portion of administrative funds. |
| V. Professional development | • Federal money will support professional development that promotes the use of "scientific, research-based and effective practice" in the classroom. |
| Motivational theory: behaviorist (educators and students will perform better when consequences are attached) | • Behaviorist |

and social studies, as well as annual assessments in reading and math from third grade through eighth grade. The nature of the assessments was not specified, except that they should allow comparisons from year to year. Samplings from the fourth-grade and eighth-grade tests of the National Assessment of Educational Progress would provide a frame of reference for interpreting state assessments.

States were required to report assessment results to parents as well as provide the public with figures showing performance by race, gender, English-language proficiency, learning disability, and socioeconomic status. (Presumably modeled after the accountability system in Texas, this feature was added to guard against disproportionate impacts on particular categories of students.)

The meat of the proposal established rewards and sanctions for school performance. States or districts were required to identify schools that have not made adequate yearly progress; schools that failed to improve after two years would be subject to corrective action, and its students could transfer to other public schools. In schools that failed to improve after three years, disadvantaged students could use Title I funds to transfer to a public or private school of their choice. States that failed to adequately narrow the achievement gap on reading and math tests could lose a portion of federal administrative funds. On the reward side of the scale, schools that narrowed the achievement gap would receive bonus funding.

Finally, federal funds would be available for professional-development activities that promoted the use of "scientific, research-based and effective practice" in classrooms.

A decade ago, the Bush proposal would have been seen as a radical attempt to assert federal control over public education. By 2001, it did little more than confirm and support existing state efforts. (The mandate for annual testing and the modest voucher proposal were new elements, but they were also the most controversial, and the administration soon signaled that it was willing to compromise on these points.)

With state and federal policy now pointed in the same direction, schools will likely have much of the next decade to make the system work. However, there *is* one potential soft spot in state and federal resolve: the willingness to apply meaningful sanctions. Thus far, only a minority of states have provided substantive consequences for underperforming schools, and even fewer have implemented those consequences. A growing number of states have also appeared skittish about linking graduation to assessments, either because of concerns about the validity of the tests or because the projected number of failures was politically unacceptable. Should educators and parents start to believe that policymakers will not follow through, the steam may go out of the accountability movement.

## Questions That Still Need To Be Answered

Beyond the policy environment, the accountability movement faces even deeper challenges. The literature reviewed in this book provides a sobering reminder of how much we *don't* know about implementing standards-based education. For accountability to have the transformative effect envisioned by policymakers, we need answers to the following questions.

1. *How realistic are the standards?* State standards vary in rigor. Some are so ambitious that educators question whether they are achievable for even a majority of children, let alone all children. Several states have had to retool their assessments because so many students were failing to meet the standards. Is this just a temporary setback that schools will overcome as they become more proficient at getting students to the desired level? Or have advocates of standards simply overreached? Standards are set through an act of professional (or political) judgment, not scientific analysis, so there is little objective evidence to reassure us that the new expectations are realistic.

2. *What must happen in classrooms for students to reach the standards?* By emphasizing motivation instead of capacity, policymakers have used a "black-box" approach that assumes a combination of clear goals and strong incentives will spur teachers to invent the necessary instructional strategies. So far the literature offers us very little evidence on how this is happening—or if it is happening. But the sophisticated thinking

demanded by many of the new standards suggests that it will not happen automatically. The ultimate success of the accountability movement may hinge on how effectively teachers are able to integrate the standards into everyday instruction—and how much help the system gives them.

3. *Does the new accountability really motivate educators and students?* The psychology is simple: reward the performers and punish the nonperformers. This carrot-and-stick motivation is so ingrained in everyday psychology that few policymakers have stopped to ask whether it will actually make a difference in school reform. So far the incentives are more rhetorical than real, but early experience suggests they get attention; when tangible consequences loom, educators and students respond actively, if somewhat defensively.

On the other hand, motivation to pass a test is not the same as motivation to achieve the standards (which is why some schools have spent their money on "test-prep" consultants instead of standards-based professional development). Increasing the stakes may simply narrow the goal to getting over the hurdle rather than integrating standards into everyday instruction.

Moreover, policymakers may have seriously misjudged teacher motivation. In a field where practitioners are driven by "psychic rewards" (Dan Lortie) and adhere to an "ethic of care" (Andy Hargreaves), promises of extra money and threats of state intervention may have limited effect. Few teachers would turn down a bonus for good performance, but there is little evidence that they will work harder—or smarter—for an extra one or two thousand dollars.

What *does* energize and excite teachers is having a visible impact on their students—not just a rise in test scores at the end of the year, but everyday evidence that students are becoming more capable and more excited about learning. If teachers find that standards-based instruction leads in that direction, then the accountability system will become a powerful motivator, with or without other incentives.

4. *How does accountability affect the rest of the classroom agenda?* Teachers and parents do not define school success solely in academic terms; they also want their children to be

happy, socially well-adjusted, and moral human beings who are capable of living a meaningful and productive life. If standards enhance that agenda, accountability will thrive. If not, the future is bleak.

As the work of Linda McNeil has shown, a purely bureaucratic approach to accountability prevents excellent teachers from doing their best and alienates students from meaningful learning. It may also have the most negative effects on students who enter school with the fewest advantages. McNeil presents us with one image of accountability's future: a disillusioned Latino student asking, "This isn't about us, is it? Doesn't the principal get a raise or something?"

The alternative image comes from Thomas Sergiovanni, who says that accountability serves a valuable purpose when it respects the "lifeworld" of schools—their beliefs, values, and passions—and when "the worth of individuals in schools is not determined by some narrow definition of effectiveness and achievement." For Sergiovanni, standards are a necessary, but not sufficient, component of quality. Only by respecting the human needs of students can schools become the kinds of institutions we want them to be. Even in an age of state mandates, school leaders must find a way to build autonomous communities of learning.

## The Leadership Challenge

In theory, accountability for student learning is shared by teachers, students, administrators, policymakers, and parents, each of whom has a role to play. In practice, when test scores are released, the hard questions will be directed at school leaders.

Few principals have been trained to manage the new accountability systems, which have clearly changed the rules. Just as technology has transformed the world of business, accountability has shaken up the expectations for educators. But just as in business, certain fundamentals are as relevant as ever. Good communication, empathy, instructional leadership, and a strong sense of ethics are still essential skills; only the issues have changed.

School leaders are still looking for ways to navigate the new environment, and research is still in the early stages, but several leadership imperatives are clear.

1. *Be the champion for standards.* Whatever their initial attitude toward standards, teachers still watch principals for signals about how seriously to take the new expectations. If leaders are blasé or dismissive, the staff will follow suit.

2. *Integrate standards into the life of the school.* Standards are not achieved by inserting new content into the curriculum. Lessons must be built around the standards; new strategies must be devised; and classroom assessment must reflect the new expectations.

3. *Give a high priority to teacher learning.* Implementing standards involves more than writing lesson plans. Teachers need opportunities to reflect deeply on the standards, their students, and their instructional repertoire, and to devise and test new strategies.

4. *Keep assessment in its place.* In theory, assessment is just feedback on student progress. In practice, the stakes are much higher, and the pressure is intense. But orienting students to pass the test, rather than achieve the standards, is ultimately self-defeating. In addition, an obsessive focus on preparing for the test can impoverish the curriculum.

5. *Devise meaningful and informative ways of reporting results.* School report cards offer a high-profile opportunity to tell the school's story and educate the public.

6. *Use data to drive the change process.* Assessment does not just tell whether things are going well or poorly—it offers clues about which strategies are working and which are not.

7. *Don't accept excuses.* Because others bear part of the responsibility for student learning—and often fail to live up to that responsibility—educators are sometimes tempted to shrug and say, "We can't do it by ourselves." Successful schools have leaders who make it clear that the undeniable barriers are no excuse for giving up and that educators should at least behave as though they could do it by themselves.

8. *Guard the lifeworld of the school.* Accountability is both a bureaucratic procedure and a moral responsibility, and the demand to achieve specified test results does not absolve

schools from meeting the other needs of students. Nothing will demoralize a staff more quickly than elevating bureaucratic requirements over their values, beliefs, and hopes for their students.

# Bibliography

Many of the items in this bibliography are indexed in ERIC's monthly catalog *Resources in Education* (*RIE*). Reports in *RIE* are indicated by an "ED" number. Journal articles, indexed in ERIC's companion catalog, *Current Index to Journals in Education,* are indicated by an "EJ" number.


Abelmann, Charles; Richard Elmore; Johanna Even; Susan Kenyon; and Joanne Marshall. *When Accountability Knocks, Will Anyone Answer?* Philadelphia, Pennsylvania: Consortium for Policy Research in Education, 1999. 59 pages. ED 428 463.

"Accountability System for Educator Preparation." Available online at www.sbec.state.tx.us/edprep/polinfo.pdf

Alderman, M. Kay. *Motivation for Achievement: Possibilities for Teaching and Learning*. Mahwah, New Jersey: Lawrence Erlbaum Associates, 1999.

American Educational Research Association. *AERA Position Statement Concerning High Stakes Testing in PreK-12 Education*. Available online at www.aera.net

American Federation of Teachers. *Making Standards Matter*. Educational Issues Policy Brief Number 11. Washington, D.C.: Author, 1999. 13 pages. ED 436 607. Available online at www.aft.org

__________. *Passing on Promotion: District Promotion Policies and Practices*. Washington, D.C.: Author, 1997. Available online at www.aft.org

__________. "Raising Student Achievement: An Internet Resource Guide for Redesigning Low-Performing Schools." Washington, D.C.: Author, 1998. Available online at www.aft.org/edissues/rsa/guide/index.htm

American Psychological Association. *Standards for Educational and Psychological Testing*. Washington, D.C.: American Psychological Association, American Educational Research Association, and National Council on Measurement in Education, 1999. 204 pages. ED 436 591.

Angaran, Joseph. "Reflection in an Age of Assessment." *Educational Leadership* 56, 6 (March 1999): 71-72. EJ 581 579.

Annenberg Institute for School Reform. "Examining Student Work." No date. Available online at www.aisr.brown.edu/accountability/lswA/index.html

__________. "A Framework for Accountability." No date. http://www.aisr.brown.edu/accountability/framework

Archbald, Douglas A. *The Reviews of State Content Standards in English Language Arts and Mathematics: A Summary and Review of Their Methods and Findings and Implications for Future Standards Development*. Washington, D.C.: National Educational Goals Panel, 1998.

Archer, Jeff. "R.I. Halts Exams in Wake of Wide-Scale Security Breaches." *Education Week* (March 17, 1999).

Aronson, Stacy R.; Sue E. Mutchler; and Diane T. Pan. *Theories of Change: Making Programs Accountable and Making Sense of Program Accountability.* Austin, Texas: Southwest Educational Development Laboratory, 1998. 69 pages. ED 426 495.

Association of Washington School Principals. "Principal Accountability Task Force Progress Report." Available online at www.awsp.org

Ball, Deborah Loewenberg. "Developing Mathematics Reform: What *Don't* We Know About Teacher Learning—But Would Make Good Working Hypotheses?" Paper prepared for conference on Teacher Enhancement in Mathematics K-6. Arlington, Virginia, November 18-20, 1994.

Ball, Deborah Loewenberg, and David K. Cohen. "Developing Practice, Developing Practitioners: Toward a Practice-Based Theory of Professional Education." In *Teaching as the Learning Profession: Handbook of Policy and Practice*, edited by Linda Darling-Hammond and Gary Sykes. 3-32. San Francisco: Jossey-Bass, 1999. ED 431 368.

Bandura, Albert A. *Self-Efficacy: The Exercise of Control*. New York: W. H. Freeman, 1997.

Barton, Paul C. *Too Much Testing of the Wrong Kind; Too Little of the Right Kind in K-12 Education*. Princeton, New Jersey: Educational Testing Service, 1999. 35 pages. ED 430 052.

Belden Russonello and Stewart. *Making the Grade: Teachers' Attitudes Toward Academic Standards and State Testing*. Washington, D.C.: Belden Russonello and Stewart, 2000. Available online at www.edweek.org (Quality Counts 2001).

Blair, Julie. "Wisconsin Legislators Approve Revised Plan for Graduation Exam." *Education Week* (October 20, 1999).

__________. "Cincinnati Teachers To Be Paid on Performance." *Education Week* (September 27, 2000).

Blase, Joseph, and Jo Roberts Blase. *Empowering Teachers: What Successful Principals Do*. Thousand Oaks, California: Corwin Press, 1994. 192 pages. ED 377 576.

Bodone, Francoise M., and Karen L. Abbie. "Teaching in a Standards-Based System: How Teachers' Voices Can Influence Policy, Preparation, and Professional Development." Paper presented at the American Educational Research Association annual meeting. Montreal, Canada, April 1999. 42 pages. ED 432 540.

Boser, Ulrich. "Pressure without Support." *Quality Counts 2001: A Better Balance: Standards, Tests, and the Tools To Succeed. Education Week* (January 10, 2001). Available online at www.edweek.org

Bracey, Gerald W. *Bail Me Out: Handling Difficult Data and Tough Questions About Schools.* Thousand Oaks, California: Corwin Press, 2000.

Bradley, Ann. "NCATE Unveils a Plan for Aspiring Elementary Teachers." *Education Week* (January 13, 1999).

_________. "New Survey Reveals Teens Yearn for High Standards." *Education Week* (February 12, 1997).

_________. "What Price Success?" *Education Week* (November 22, 1995).

Brandt, Richard M. *Incentive Pay and Career Ladders for Today's Teachers: A Study of Current Programs and Practices.* Albany, New York: State University of New York Press, 1990.

Brooks, Sarah R. *How States Can Hold Schools Accountable: The Strong Schools Model of Standards-Based Reform.* Seattle, Washington: Center on Reinventing Public Education, 2000.

Buchmann, Margaret. "Teaching Knowledge: The Lights That Teachers Live By." *Oxford Review of Education* 13, 2 (1987): 151-64. EJ 358 532.

Burger, Don. "Designing a Sustainable Standards-Based Assessment System." In *What's Noteworthy on Learners, Learning and Schooling.* Edited by Sandra Carroll Berger and Diane McIntyre Wilber. Aurora, Colorado: Mid-Continent Regional Laboratory, 1995. 71 pages. ED 398 641.

Business Coalition for Education Reform. "Making Academics Count." Available online at www.bcer.org

Calkins, Lucy; Kate Montgomery; and Donna Santman. "Helping Children Master the Tricks and Avoid the Traps of Standardized Tests." *Practical Assessment, Research and Evaluation* 6, 8 (1999). ED 429 987. Available online at http://ericae.net/pare/getvn.asp?v=6&n=8

Chan, Yvonne; Guillermo Galarza; Suzanne Llamas; Eileen Kellor; and Allan Odden. *A Case Study of the Vaughn Next Century Learning Center's School-Based Performance Award Program.* Madison, Wisconsin: Consortium for Policy Research in Education, 1999.

Clotfelter, Charles T., and Helen F. Ladd. "Recognizing and Rewarding Success in Public Schools." In *Holding Schools Accountable: Performance-Based Reform in Education,* edited by Helen F. Ladd. 23-63. Washington, D.C.: The Brookings Institution, 1996. 382 pages. ED 396 428.

Cohen, David K. "Rewarding Teachers for Student Performance." In *Rewards and Reform: Creating Educational Incentives That Work,* edited by Susan H. Fuhrman and Jennifer O'Day. 60-112. San Francisco: Jossey-Bass, 1997.

Consortium for Policy Research in Education. "Developing Content Standards: Creating a Process for Change." *CPRE Policy Briefs.* RB-10. New

Brunswick, New Jersey: Consortium for Policy Research in Education, 1993. 9 pages. ED 362 981.

_________. "Repeating Grades in School: Current Practice and Research Evidence." *CPRE Policy Briefs*. RB-04. January 1990. 9 pages. ED 323 585.

Coronado Village Elementary School. "School Accountability Report Card." Coronado Village, California: 1998. Available online at www.village.coronado.k12.ca.us/report/print.html

Cort, Julie. "Test Cheating Scandal Leaves Town Unable To Fill in the Blanks." *Los Angeles Times* (August 10, 2000).

Csikszentmihalyi, Mihaly. *Flow: The Psychology of Optimal Experience*. New York: Harper Perennial, 1990.

Dalton, Stephanie S. *Pedagogy Matters: Standards for Effective Teaching*. Research Report 4. Washington D.C.: Center for Research on Education, Diversity and Excellence, 1998. 49 pages. ED 424 769.

Darling-Hammond, Linda. "Accountability for Professional Practice." *Teachers College Record* 91, 1 (Fall 1989): 59-80. EJ 398 426.

Darling-Hammond, Linda, and Deborah Loewenberg Ball. *Teaching for High Standards: What Policymakers Need To Know and Be Able to Do*. Philadelphia: Center for Policy Research in Education, 1998. 40 pages. ED 426 491.

Darling-Hammond, Linda, and Milbrey W. McLaughlin. "Policies That Support Professional Development in an Era of Reform." In *Teacher Learning: New Policies, New Practices*, edited by Milbrey W. McLaughlin and Ida Oberman. 202-18. New York: Teachers College Press, 1996. 252 pages. ED 398 183.

Deci, Edward L. *Why We Do What We Do*. New York: G. P. Putnam's Sons, 1995.

Dell'Angela, Tracy. "Aurora Teachers Feeling New Policy's Heat." *Chicago Tribune* (October 2, 2000).

Denver Public Schools. "Denver Public Schools Pay for Performance Plan." Available online at www.dpsnet.denver.k12.co.us/news/documents/PayforPerformance.html

Education Commission of the States. *Governing America's Schools: Changing the Rules*. Denver, Colorado: Author, 1999. 76 pages. ED 439 513.

_________. *State Takeovers and Reconstitutions*. Denver, Colorado: Author, 2000.

Eisner, Elliot W. "The Uses and Limits of Performance Assessment." *Phi Delta Kappan* 80, 9 (May 1999): 658-60. EJ 585 676.

Elmore, Richard F., and Susan H. Fuhrman. "Opportunity-to-Learn Standards and the State Role in Education." *Teachers College Record* 96, 3 (Spring 1995): 432-57. EJ 505 807.

Ervin, Keith. "Only 1 in 4 Pass Entire 4th-Grade State Test." *Seattle Times* (October 21, 2000).

Farkas, Steve; Jean Johnson; and Tony Faleno. *On Thin Ice: How Advocates and Opponents Can Misread the Public's Views on Vouchers and Charter Schools*. New York: Public Agenda, 1999.

Finn, Chester. "Who's Afraid of the Big Bad Test?" In *Debating the Future of American Education: Do We Need National Standards and Assessments?*, edited by Diane Ravitch. 120-44. Washington D.C.: The Brookings Institution, 1995. ED 387 917.

Finn, Chester E., Jr., and Michael J. Petrilli (Eds.). *The State of State Standards 2000*. Washington, D.C.: The Thomas B. Fordham Foundation, 2000. 183 pages. ED 439 133. Available online at http://edexcellence.net

Finn, Chester E., Jr., and Danielle Dunne Wilcox. "Teachers Should Be Graded on How Well Their Students Are Learning." *Los Angeles Times* (January 13, 2000).

Foriska, Terry J. *Restructuring Around Standards: A Practitioner's Guide to Design and Implementation*. Thousand Oaks, California: Corwin press, 1998. 147 pages. ED 422 642.

Galley, Michelle. "Chicago To Size Up Parents with 'Checklist'." *Education Week* (May 31, 2000).

Gibson, Sherri, and Myron Dembo. "Teacher Efficacy: A Construct Validation." *Journal of Educational Psychology* 76, 4 (August 1984): 568-82. EJ 306 050.

Glass, Don. *The Cart Before the Horse Before the Cart: How Deeper Understandings of Standards, Instruction, and Assessment Emerge from Examining Student Work*. Providence, Rhode Island: Annenberg Institute for School Reform, 2000. Available online at www.aisr.brown.edu

Gough, Pauline. "Moving Toward the Danger." *Phi Delta Kappan* 81, 8 (April 2000): 3-4.

Grant, Jim, and Bob Johnson. "Overcoming Obstacles to High Standards." *Principal* 77, 2 (November 1997): 34-36. EJ 553 794.

Haney, Walt. "The Myth of the Texas Miracle in Education." *Education Policy Analysis Archives* 8, 41 (August 2000). Available online at http://epaa.asu.edu/epaa/v8n41

Hardy, Roy. "Performance Assessments: Examining the Costs." In *Implementing Performance Assessment: Promises, Problems, and Challenges*, edited by Michael B. Kane and Ruth Mitchell. 107-17. Mahwah, New Jersey: Erlbaum, 1996. ED 433 344.

Hargreaves, Andy. *Changing Teachers, Changing Times: Teachers' Work and Culture in the Postmodern Age*. New York: Teachers College Press, 1994.

Harris, Douglas, and Judy Carr. *How to Use Standards in the Classroom*. Alexandria, Virginia: Association for Supervision and Curriculum Development, 1996. 75 pages. ED 400 598.

Hawley, Willis D., and Linda Valli. "The Essentials of Effective Professional Development: A New Consensus." In *Teaching as the Learning Profession:*

*Handbook of Policy and Practice,* edited by Linda Darling-Hammond and Gary Sykes. 127-50. San Francisco: Jossey-Bass, 1999. 426 pages. ED 431 368.

Hoff, David. "NAEP Weighed as Measure of Accountability." *Education Week* (March 8, 2000).

_________. "Standards at Crossroads After Decade." *Education Week* 19, 3 (September 22, 1999): 1, 9.

Hord, Shirley M. *Professional Learning Communities: Communities of Continuous Inquiry and Improvement.* Austin, Texas: Southwest Educational Development Laboratory, 1997. 71 pages. ED 410 659. Available online at www.sedl.org

Horner, Matina. "Toward an Understanding of Achievement-Related Conflicts in Women." *Journal of Social Issues* 28, 2 (Spring 1972): 157-76. EJ 065 572.

Huff, Darrell. *How To Lie with Statistics.* New York: W.W. Norton, 1954.

Jackson, Philip W. *Life in Classrooms.* New York: Holt, Rinehart and Winston, 1968.

Jacobs, Heidi Hayes. *Mapping the Big Picture: Integrating Curriculum and Assessment K-12.* Alexandria, Virginia: Association for Supervision and Curriculum Development, 1997. 147 pages. ED 411 323.

Jamentz, Kate. "Authentic Accountability." *Thrust for Educational Leadership* (January 1998a).

_________. *Standards: From Document to Dialogue.* San Francisco: WestEd, 1998b. 82 pages. ED 423 599.

Johnson, Henry; Elsie Leak; Gary Williamson; Eileen Kellor; Tony Milanowski; Allan Odden; and Jason Hanna. *A Case Study of the State of North Carolina's School-Based Performance Award Program.* Madison, Wisconsin: Consortium for Policy Research in Education, 1999. Available online at www.wcer.wisc.edu/cpre

Johnson, Jean, and Ann Duffett. *Standards and Accountability: Where the Public Stands.* New York: Public Agenda, 1999.

Johnston, Robert C. "Texas Presses Districts in Alleged Test-Tampering Cases." *Education Week* (March 17, 1999).

Johnston, Robert C., and Debra Viadero. "Unmet Promise: Raising Minority Achievement." *Education Week* (March 15, 2000).

Jones, Bryan L. "Me, Ms. DeLuge, and the Macarena." *Teacher Magazine* (January 2001). Available online at www.edweek.org

Jones, M. Gail; Brett D. Jones; Belinda Hardin; Lisa Chapman; Tracie Yarborough; and Marcia Davis. "The Impact of High-Stakes Testing on Teachers and Students in North Carolina." *Phi Delta Kappan* (November 1999): 199-203. EJ 597 091.

Joyce, Bruce, and Emily Calhoun. "School Renewal: An Inquiry, Not a Formula." *Educational Leadership* 52, 7 (April 1995): 51-55. EJ 502 912.

Keeney, Lorraine. *Using Data for School Improvement*. Providence, Rhode Island: Annenberg Institute for School Reform, 1998.

Keller, Bess. "Incentives for Test-Takers Run the Gamut." *Education Week* (May 3, 2000).

Keller, John M. "Motivational Systems." In *Handbook of Human Performance Technology*, 2nd editon, edited by Harold D. Stolovitch and Erica J. Keeps. 373-94. San Francisco: Jossey-Bass Pfeiffer, 1999.

Kelley, Carolyn; Allan Odden; Anthony Milanowski; and Herbert Heneman III. "The Motivational Effects of School-Based Performance Awards." *CPRE Policy Briefs*. RB-29. February 2000. 13 pages. ED 440 473.

Kiplinger, Vonda L., and Robert L. Linn. *Raising the Stakes of Test Achievement: The Impact on Student Performance on the NAEP.* Los Angeles: National Center for Research on Evaluation, Standards, and Student Testing, 1993. 76 pages. ED 378 221.

Kohn, Alfie. *Punished by Rewards: The Trouble with Gold Stars, Incentive Plans, A's, Praise, and Other Bribes*. Boston: Houghton Mifflin, 1993.

Kronholtz, June. "Colorado School Districts Promise To Make the Grade." *Wall Street Journal* (October 29 1999).

Lachat, Mary Ann. *Standards, Equity and Cultural Diversity*. Providence, Rhode Island: Northeast and Islands Regional Educational Laboratory, 1999.

Lake, Robin; Maria McCarthy; Sara Taggart; and Mary Beth Celio. *Making Standards Stick: A Followup Look at Washington State's School Improvement Efforts in 1999-2000*. Seattle: Center on Reinventing Public Education, 2000.

Lake, Robin; Paul T. Hill; Lauren O'Toole; and Mary Beth Celio. *Making Standards Work: Active Voices, Focused Learning*. Seattle: Center on Reinventing Public Education, 1999. 22 pages. ED 430 268.

Lankard, Bettina. "Business/Industry Standards and Vocational Program Accountability." *ERIC Digest* 157. Columbus, Ohio: Clearinghouse on Adult, Career, and Vocational Education, 1995. 4 pages. ED 383 857.

Learning First Alliance. *Standards and Accountability: A Call by the Learning First Alliance for Mid-Course Corrections*. Washington, D.C.: Learning First Alliance, 2001. Available online at www.learningfirst.org

Lee, Valerie E; Julia B. Smith; Tamara E. Perry; and Mark A. Smylie. *Social Support, Academic Press, and Student Achievement: A View from the Middle Grades in Chicago*. Chicago: Consortium on Chicago School Research, 1999. 41 pages. ED 439 213.

Lepper, Mark R. "Motivational Considerations in the Study of Instruction." *Cognition and Instruction* 5, 4 (1998): 289-309.

Levesque, Karen; Denise Bradby; and Kristi Rossi. "Using Data for Program Improvement: How Do We Encourage Schools To Do It?" *CenterFocus* 12 (May 1996). ED 392 983. Available online at www.vocserve.berkeley.edu/CenterFocus/CF12.html

Levesque, Karen; Denise Bradby; Kristi Rossi; and Peter Teitelbaum. *At Your Fingertops: Using Everyday Data To Improve Schools*. Berkeley, California: MPR Associates, 1998.

Lieberman, Ann. "Practices That Support Teacher Development: Transforming Conceptions of Teacher Learning." In *Teacher Learning: New Policies, New Practices*, edited by Milbrey W. McLaughlin and Ida Oberman. 185-201. New York: Teacher College Press, 1996. 252 pages. ED 398 183.

Lieberman, Ann, and Milbrey W. McLaughlin. "Networks for Educational Change: Powerful andf Problematic." In *Teacher Learning: New Policies, New Practices*, edited by Milbrey W. McLaughlin and Ida Oberman. 63-72. New York: Teacher College Press, 1996. 252 pages. ED 398 183.

Linn, Robert L. "Assessments and Accountability." *Educational Researcher* 29, 2 (March 2000): 4-16. EJ 602 768.

Linn, Robert L., and Eva L. Baker. "Back to Basics—Indicators as a System." *The CRESST Line*. (Winter 1998): 1-3. Available online at http://cresst96.cse.ucla.edu/CRESST/pages/newsletters.htm

Little, Judith Warren. "Organizing Schools for Teacher Learning." In *Teaching as the Learning Profession: Handbook of Policy and Practice*, edited by Linda Darling-Hammond and Gary Sykes. 233-62. San Francisco: Jossey-Bass, 1999. ED 431 368.

Lortie, Dan C. *Schoolteacher: A Sociological Study*. Chicago: The University of Chicago Press, 1975.

Louis, K. S., and S. D. Kruse. *Professionalism and Community: Perspectives on Reforming Urban Schools*. Thousand Oaks, California: Corwin Press, 1995

Lumsden, Linda. *Student Motivation: Cultivating a Love of Learning*. Eugene, Oregon: ERIC Clearinghouse on Educational Management, University of Oregon, 1999. 114 pages.

Malone, Thomas. "Toward a Theory of Intrinsically Motivating Instruction." *Cognitive Science* 4 (1981) 333-70.

Marks, Helen M., and Karen Seashore Louis. "Teacher Empowerment and the Capacity for Organizational Learning." *Educational Administration Quarterly* 35 (December 1999 Supplemental): 707-51.

Marzano, Robert J., and John S. Kendall. *Awash in a Sea of Standards*. Aurora, Colorado: Mid-Continent Regional Educational Laboratory, 1998. Available online at www.mcrel.org

__________. *A Comprehensive Guide to Designing Standards-Based Districts, Schools, and Classrooms*. Alexandria, Virginia: Association for Supervision and Curriculum Development, 1996. 318 pages. ED 414 625.

__________. *Content Knowledge: A Compendium of Standards and Benchmarks for K-12 Education*. 2nd Edition. Aurora, Colorado: Mid-Continent Regional Educational Laboratory, 1997.

Marzano, Robert J.; John S. Kendall; and Louis F. Cicchinelli. *What Americans Believe Students Should Know: A Survey of U.S. Adults*. Aurora, Colorado: Mid-Continent Regional Educational Laboratory, 1998. 230 pages. ED 426 500. Available online at www.mcrel.org

Maslow, Abraham. *Toward a Psychology of Being*. 3rd Edition. Princeton, New Jersey: Van Nostrand, 1968.

Mathews, Jay. "When Parents and Students Grade Staff." *School Administrator* (October 2000).

May, Heather. "Educators: Using Tests To Rank Schools Will Create Cheaters, Dropouts." *Salt Lake City Tribune* (August 1, 2000).

McDonnell, Lorraine M. *Policymakers' Views of Student Assessment*. Los Angeles: National Center for Research on Evaluation, Standards, and Student Assessment, 1994.

McKenzie, Richard B., and Dwight R. Lee. *Managing Through Incentives: How To Develop a More Collaborative, Productive, and Profitable Organization*. New York: Oxford University Press, 1998.

McNeil, Linda. *Contradictions of School Reform: Educational Costs of Standardized Testing*. New York: Routledge, 2000.

Mehrens, William A. ""Consequences of Assessment: What Is the Evidence?" *Education Policy Analysis Archives* 6, 13 (July 1998). EJ 582 965. Available online at http://olam.ed.asu.edu/epaa/v6n13.html

Mitchell, Ruth. *Front-End Alignment: Using Standards To Steer Educational Change*. Washington, D.C.: The Education Trust, 1996.

Mohrmann, Susan Albers, and Edward E. Lawler III. "Motivation for School Reform." In *Rewards and Reform: Creating Educational Incentives That Work*, edited by Susan H. Fuhrman and Jennifer O'Day. 115-143. San Francisco: Jossey-Bass, 1997.

Moore, Kathleen. *State-Mandated Educational Assessments: Survey of Parents' Perceptions*. No location: Harris Interactive, 2000. Available online at www.ascd.org

National Board of Professional Teaching Standards. *What Teachers Should Know and Be Able To Do*. 1989. Available online at www.nbpts.org.

National Center for History in the Schools. *National Standards for History*. Basic Edition. 1996. Available online at www.sscnet.ucla.edu

National Commission on Excellence in Education. *A Nation at Risk: A Report to the Nation and the Secretary*. Washington, D.C.: Government Printing Office, 1983.

National Council of Teachers of English. *Standards for the English Language Arts*. Urbana, Illinois: National Council of Teachers of English, 1996. Available online at www.ncte.org

National Council of Teachers of Mathematics. *Principles and Standards for School Mathematics*. Reston, Virginia: National Council of Teachers of Mathematics, 2000. Available online at www.nctm.org

National Education Commission on Time and Learning. *Prisoners of Time*. Washington, D.C.: Government Printing Office, 1994. 59 pages. ED 366 115.

National Educational Goals Panel. *Building a Nation of Learners*. Washington, D.C.: National Educational Goals Panel, 1999. Available online at www.negp.gov

National Forum on Assessment. *Principles and Indicators for Student Assessment Systems*. Cambridge, Massachusetts: FairTest, 1995. 38 pages. ED 400 334. Available online at www.fairtest.org

National Foundation for the Improvement of Education. *Creating Teacher-Led Professional Development Centers*. Washington, D.C.: National Foundation for the Improvement of Education, 2000. http://www.nfie.org

National Research Council. *High Stakes: Testing for Tracking, Promotion, and Graduation*. Washington, D.C.: National Research Council, 1999.

Natriello, Gary, and Aaron M. Pallas. *The Development and Impact of High Stakes Testing*. Teachers College, Columbia University and Michigan State University, 1998. Available online at www.columbia.edu/~gjn6/histake.html

Neill, Monty. *Testing Our Children: A Report Card on State Assessment Systems*. Cambridge, Massachusetts: FairTest, 1997. Available online at www.fairtest.org

Nelson-Barber, Sharon. "A Better Education for Every Child: The Dilemma for Teachers of Culturally and Linguistically Diverse Students." In *Including Culturally and Linguistically Diverse Students in Standards-Based Reform: A Report on McREL's Diversity Roundtable I*. Aurora, Colorado: Mid-Continent Regional Educational Laboratory, 1999. Available online at www.mcrel.org

Newmann, Fred M.; M. Bruce King; and Mark Rigdon. *Accountability and School Performance: Implications from Restructuring Schools*. Madison, Wisconsin: Center on Organization and Restructuring of Schools, 1996. 70 pages. ED 412 631.

New York State Education Department. Sample report cards can be found online at www.emsc.nysed.gov/repcrd2000

Oakes, Jeannie. *Educational Indicators: A Guide for Policymakers*. Center for Policy Research in Education, 1986. 49 pages. ED 315 920.

O'Day, Jennifer. "Introduction: Incentives and School Improvement." In *Rewards and Reform: Creating Educational Incentives That Work*, edited by Susan H. Fuhrman and Jennifer O'Day. 1-16. San Francisco: Jossey-Bass, 1997.

O'Day, Jennifer; Margaret E. Goertz; and Robert E. Floden. "Building Capacity for Education Reform." *CPRE Policy Briefs*. RB-18. (December 1995).

Odden, Allan. "Incentives, School Organization, and Teacher Compensation." In *Rewards and Reform: Creating Educational Incentives That Work*, edited by Susan H. Fuhrman and Jennifer O'Day. 226-56. San Francisco: Jossey-Bass, 1997.

Odden, Allan, and Carolyn Kelley. *Paying Teachers for What They Know and Do: New and Smarter Compensation Strategies To Improve Schools*. Thousand Oaks, California: Corwin Press, 1997. 204 pages. ED 404 312.

Office of Technology Assessment. *Testing in American Schools: Asking the Right Questions*. (OTA-SET-519). Washington, D.C.: U.S. Government Printing Office, 1992. 314 pages. ED 340 770.

Olson, Lynn. "Finding the Right Mix." In *Quality Counts 2001: A Better Balance: Standards, Tests, and the Tools To Succeed*. 12-15, 17-20. Bethesda, Maryland: Editorial Projects in Education, Inc., 2001.

_________. "Indiana Case Focuses on Special Ed." *Education Week* (May 31, 2000a).

_________. "Making Every Test Count." In *Quality Counts '99: Rewarding Results, Punishing Failure*. 11, 15-16, 18-20. Bethesda, Maryland: Editorial Projects in Education, Inc., 1999a.

_________. "Report Cards for Schools." In *Quality Counts '99: Rewarding Results, Punishing Failure*. 217-28. 32-36. Bethesda, Maryland: Editorial Projects in Education, Inc., 1999b.

_________. "Shining a Spotlight on Results." In *Quality Counts '99: Rewarding Results, Punishing Failure*. 8-10. Bethesda, Maryland: Editorial Projects in Education, Inc., 1999c.

_________. "States Ponder New Form of Diploma." *Education Week* (June 21, 2000b).

Osborne, Peter, and Diane Plastrik. *The Reinventor's Fieldbook: Tools for Transforming Your Government*. San Francisco: Jossey-Bass, 2000.

Phi Delta Kappa. "International Curriculum Management Audit Center." Available online at www.pdkintl.org/profdev/icmac/whatis.htm

Pinder, Craig. *Work Motivation in Organizational Behavior*. Upper Saddle River, New Jersey: Prentice Hall, 1998.

Popham, W. James. "Why Standardized Tests Don't Measure Educational Quality." *Educational Leadership* 56, 6 (March 1999): 8-15. EJ 581 564.

Porter, Andrew, "The Use and Misuses of Opportunity-to-Learn Standards." In *Debating the Future of American Education: Do We Need National Standards and Assessments?*, edited by Diane Ravitch. Washington, D.C.: The Brookings Institution, 1995. 182 pages. ED 387 917.

Portner, Jessica. "Most Virginia Schools Fall Short on High-Stakes Tests." *Education Week* 19, 1 (September 8 1999): 16.

Powell, Arthur G. "Motivating Students To Learn: An American Dilemma." In *Rewards and Reform: Creating Educational Incentives That Work*, edited by Susan H. Fuhrman and Jennifer O'Day. 19-59. San Francisco: Jossey-Bass, 1996. 365 pages. ED 426 453.

Pritchard, Ivor. "Judging Standards in Standards-Based Reform." *Perspective* 8, 1 (Summer 1996). 26 pages. ED 412 302.

Public Agenda. "Reality Check 2000." *Education Week* (February 16, 2000).

Rallis, Sharon F., and Margaret M. MacMullen. "Inquiry-Minded Schools: Opening Doors for Accountability." *Phi Delta Kappan* 81, 10 (June 2000): 766-73.

Ratnesar, Romesh. "The Homework Ate My Family." *Time* 153, 3 (January 25 1999): 54-63. EJ 582 477.

Ravitch, Diane. *National Standards in Education: A Citizen's Guide*. Washington, D.C.: The Brookings Institution, 1995.

Raywid, Mary Anne. "Finding Time for Collaboration." *Educational Leadership* 51, 1 (September 1993): 30-34. EJ 468 684.

Regents of the University of California. *School Report Card*. Los Angeles: University of California, National Center for Research on Evaluation, Standards, and Student Testing (CRESST), 1999.

Resnick, Lauren, and Katherine Nolan. "Standards for Education." In *Debating the Future of American Education: Do We Need National Standards and Assessments?*, edited by Diane Ravitch. Washington, D.C.: The Brookings Institution, 1995. 182 pages. ED 387 917.

Rhode Island Education Department. *Rhode Island Beginning Teacher Standards*. Third edition. 1999. Available online at www.ridoe.net/teacher_cert/teacher_prep/pro_standards.htm

Roderick, Melissa; Anthony S. Bryk; Brian A. Jacob; John Q. Easton; and Elaine Allensworth. *Ending Social Promotion: Results from the First Two Years*. Chicago: Consortium on Chicago School Research, 1999. 79 pages. ED 439 214. Available online at www.consortium-chicago.org

Rose, Lowell C., and Alex Gallup. "The 31st Annual Phi Delta Kappan/Gallup Poll of the Public's Attitudes Toward the Public Schools." *Phi Delta Kappan* 81, 1 (September 1999): 41-56. EJ 592 905.

Sacks, Peter. *Standardized Minds: The High Price of America's Testing Culture and What We Can Do To Change It*. Cambridge, Massachusetts: Perseus Books, 2000.

Sanders, William L. "Value-Added Assessment." *School Administrator* 11, 55 (December 1998): 24-27. EJ 577 201.

Sandham, Jessica L. "Florida OKs First Statewide Voucher Plan." *Education Week* 18, 34 (May 5, 1999): 1, 21.

Schlechty, Phillip. *Schools for the Twenty-First Century: Leadership Imperatives for Educational Reform*. San Francisco: Jossey-Bass, 1990.

Schmoker, Mike, and Robert Marzano. "Realizing the Promise of Standards-Based Education." *Educational Leadership* 56, 6 (March 1999) 17-21. EJ 581 565.

Schrag, Peter. "High Stakes Are for Tomatoes." *Atlantic Monthly* 286, 2 (August 2000): 19-21.

Secretary's Commission on Achieving Necessary Skills (SCANS). *What Work Requires of Schools: SCANS Report for America 2000*. Washington, D.C.: U.S. Department of Labor, 1991. 60 pages. ED 332 054.

Sergiovanni, Thomas J. *The Lifeworld of Leadership: Creating Culture, Community, and Personal Meaning in Our Schools*. San Francisco: Jossey-Bass, 2000. 218 pages. ED 436 829.

Serow, Robert. "Called To Teach: A Study of Highly Motivated Preservice Teachers." *Journal of Research and Development in Education* 27, 2 (Winter 1994): 65-72. EJ 488 769.

Shavelson, Richard J.; Lorraine McDonnell; and Jennie Oakes. "What Are Educational Indicators and Indicator Systems?" *ERIC Digest*. Washington, D.C.: ERIC Clearinghouse on Assessment and Evaluation, 1991a. 4 pages. ED 338 701.

__________. "Steps in Designing an Indicator System." *Practical Assessment, Research and Evaluation* 2, 12 (1991b). 4 pages. ED 338 700. Available online at http://ericae.net

Skinner, B. F. *Beyond Freedom and Dignity*. New York: Bantam/Vintage, 1972.

Smith, Nelson. *Standards Mean Business*. Washington, D.C.: National Alliance of Business, 1996. Available online at www.nab.com

Solmon, Lewis C., and Michael Podgursky. *The Pros and Cons of Performance-Based Compensation*. No Location: Milken Family Foundation, 2000. Available online at www.mff.org

Sommerfeld, Meg. "Employers Discount School Factors in Hiring Decisions." *Education Week* (March 1, 1995).

Southern Regional Education Board. *Getting Results: A Fresh Look at School Accountability*. Atlanta: Southern Regional Education Board, 1998. 31 pages. ED 426 510.

__________. *Getting Results with Accountability: Rating Schools, Assisting Schools, Improving Schools*. Atlanta: Southern Regional Education Board, 2000.

Sparks, Dennis, and Stephanie Hirsch. *A New Vision for Staff Development*. Alexandria, Virginia: Association for Supervision and Curriculum Development, 1997. 115 pages. ED 410 201.

Spillane, James P. *District Leaders' Perceptions of Teacher Learning*. Philadelphia: Consortium for Policy Research in Education, 2000.

Stager, Gary. "Against Educational Standards? An Exclusive Interview with Susan Ohanian." *Curriculum Administrator* 35, 4 (April 1999): 20, 22-23. EJ 595 970.

State Board for Educator Preparation. "Accountability System for Educator Preparation." 1999. Available online at http://www.sbec.state.tx.us/edprep/edprep.htm

Steinberg, Jacques. "Blue Books Closed, Students Boycott Standardized Tests." *New York Times* (April 13, 2000).

Stipek, Deborah. *Motivation to Learn: From Theory to Practice.* Boston: Allyn and Bacon, 1998. 292 pages. ED 369 773.

Sykes, Gary. "Teacher and Student Learning: Strengthening Their Connection." In *Teaching as the Learning Profession: Handbook of Policy and Practice,* edited by Linda Darling-Hammond and Gary Sykes. 151-79. San Francisco: Jossey-Bass, 1999. ED 431 368.

Teicher, Stacy A. "In This AP Program, It Pays To Study." *Christian Science Monitor* (October 17, 2000).

Tell, Christine A.; Francoise Bodone; and Karen L. Addie. "Teaching in a Standards-Based System: How Teachers' Voices Can Influence Policy, Preparation, and Professional Development." Paper presented at the American Educational Research Association annual meeting, Montreal, Canada, April 23, 1999.

Thompson, Charles L., and John S. Zeuli. "The Frame and the Tapestry: Standards-Based Reform and Professional Development." In *Teaching as the Learning Profession: Handbook of Policy and Practice,* edited by Linda Darling-Hammond and Gary Sykes. 341-75. San Francisco: Jossey-Bass, 1999. ED 431 368.

United States Department of Education. *Taking Responsibility for Ending Social Promotion: A Guide for Educators and State and Local Leaders.* Washington, D.C.: U.S. Department of Education, 1999. 92 pages. ED 430 319. Available online at www.ed.gov/pubs/socialpromotion

__________. *Turning Around Low-Performing Schools: A Guide for State and Local Leaders.* May 1998. 90 pages. ED 419 301. Available online at www.ed.gov/pubs/turning

Viadero, Debra. "Testing System in Texas Yet To Get Final Grade." *Education Week* (May 31, 2000).

Viadero, Debra, and Robert C. Johnston. "Lifting Minority Achievement: Complex Answers." *Education Week* (April 5, 2000).

Wagner, Robert B. *Accountability in Education: A Philosophical Inquiry.* New York: Routledge, 1989.

Wald, Penelope J., and Michael S. Castleberry (Eds.). *Educators as Learners: Creating a Professional Learning Community in Your School.* Alexandria, Virginia: Association for Supervision and Curriculum Development, 2000. 140 pages. ED 439 099.

Washington State Commission on Student Learning. *Essential Academic Learning Requirements Technical Manual.* Olympia, Washington: Washington State Commission on Student Learning, 1998. Available online at www.k12.wa.us

Wasley Patricia. "The Practical World of Teacher Leaders: Assumptions, Attitudes and Acrophobia." In *Staff Development for Education in the '90s: New Demands, New Realities, New Perspectives*, edited by Ann Lieberman and Lynne Miller. 158-83. New York: Teachers College Press, 1991.

Watts, James A.; Gale F. Gaines; and Joseph D. Creech. *Getting Results: A Fresh Look at School Accountability*. Atlanta: Southern Regional Education Board, 1998. 31 pages. ED 426 510.

Willis, Scott. "The Accountability Question." *Education Update* 41, 7 (November 1999). Available online at www.ascd.org/readingroom/edupdate/1999/1nov.html

Willis, Tom; Kyna Koch; Gretchen Lampe; Robert Young; Eileen Kellor; and Allen Odden. *A Case Study of the State of Kentucky's School-Based Performance Award Program*. Madison, Wisconsin: Consortium for Policy Research in Education, 1999. Available online at www.wcer.wisc.edu/cpre

Wolf, Dennie Palmer, and Anne Marie White. "Charting the Course of Student Growth." *Educational Leadership* 57, 5 (February 2000).

Zeidner, Moshe. *Test Anxiety: The State of the Art*. New York: Plenum Press, 1998.

# Subject Index

# Author Index

## M



## N



## O



## P

## R



## S



## T



## U



## V



## W



## Y



## Z

# Publications

*Available from the ERIC Clearinghouse on Educational Management*

**The New Standards & Accountability: *Will Rewards and Sanctions Motivate America's Schools to Peak Performance?***

Larry Lashway • 2001 • 6x9 inches • viii + 216 pages • perfect bind • ISBN 0-86552-149-2 • $14.00. Code: EMONSA

**In Search of More Productive Schools: A Guide to Resource Allocation in Education**

Lawrence O. Picus • 2001 • 6x9 inches • xiii + 122 pages • perfect bind • ISBN 0-86552-147-6 • $12.75. Code: EMOSMP

**Safe School Design: A Handbook for Educational Leaders *Applying the Principles of Crime Prevention Through Environmental Design***

Tod Schneider, Hill Walker, Jeffrey Sprague • 2000 • 8 $^1/_2$ x 11 inches • xvi + 96 pages • perfect bind • ISBN 0-86552-148-4 • $18.00. Code: EMOSSD

**Learning Experiences in School Renewal: An Exploration of Five Successful Programs**

Bruce Joyce and Emily Calhoun • 1996 • 6 x 9 inches • viii + 208 pages • perfect bind • ISBN: 0-86552-133-6 • $14.50. Code: EMOLES

**Measuring Leadership: A Guide to Assessment for Development of School Executives**

Larry Lashway • 1999 • 6x9 inches • viii + 117 pages • perfect bind • ISBN 0-86552-140-9 • $9.75. Code: EMOMLG

**Roadmap to Restructuring: Charting the Course of Change in American Education**

David T. Conley • Second Edition • 1997 • 6 x 9 inches • xvi + 571 pages • Cloth ISBN: 0-86552-136-0 ($34.95) Code: EMORMC • Paper ISBN 0-86552-137-9 ($23.95) Code: EMORMP

**Leading with Vision**

Larry Lashway • 1997 • 6 x 9 inches • xii + 148 pages • perfect bind • ISBN: 0-86552-138-7 • $13.50. Code: EMOLWV

**School Leadership: Handbook for Excellence**

Edited by Stuart C. Smith and Philip K. Piele • Third Edition • 1997 • xvi + 432 pages • Cloth ISBN 0-86552-134-4 ($29.95) Code: EMOSLC • Paper ISBN 0-86552-135-2 ($19.95) Code: EMOSLP

**Student Motivation: Cultivating a Love of Learning**

Linda Lumsden • 1999 • 6x9 inches • vi + 113 pages • perfect bind • ISBN 0-86552-141-7 • $9.50. Code: EMOSMC

**How to Order:** You may place an order by sending a check or money order, mailing or faxing a purchase order, or calling with a Visa or MasterCard number. Add 10% for S&H (minimum $4.00). Make payment to University of Oregon/ERIC. Shipping is by UPS ground or equivalent.

Telephone (800) 438-8841 • Fax (541) 346-2334

Publications Sales
ERIC Clearinghouse on Educational Management
5207 University of Oregon
Eugene, OR 97403-5207

You can also order online (with Visa or MasterCard) from our website—your gateway to information about educational policy and management.

http://eric.uoregon.edu

**http://eric.uoregon.edu**

## You Can Find Everything From Here

Dissemination of information is our mission, and the World Wide Web is a primary means for dispensing this information to our users.

### Online Directory of Organizations in Educational Management

Guiding you to sources of information on policy and management of elementary and secondary schools, this online Directory includes profiles of more than 160 organizations. The Directory supplies names, addresses, telephone and fax numbers, missions, subject areas, publication topics, and services, all continually updated through ongoing contact with the organizations.

### Hot Topics

Looking for resources on class size, educational vouchers, charter schools, school safety, leadership, whole-school reform, and other topics of current interest?

### Résumés of In-Process Documents

You can search up to 1,200 résumés of recently processed, cutting-edge documents and journal articles before they appear in the ERIC database.

### Trends and Issues

At this ever-changing section of the website, you can scan summaries and analyses of major issues, trends, and innovations in educational management.

Among the most popular of the fifteen issues currently available are Role of the School Leader, School Choice, Social and Economic Context, Educational Governance, and School Organization.

### Links to Other Resources

We maintain a carefully selected list of links to other organizations and resources, especially within the U.S. Department of Education.

There are also scores of links to publishers of journals and books in educational management.

### Online Publications Catalog

You can browse or search our catalog, which includes detailed information on every one of our publications.

You can also order publications online through our secure server using Visa and MasterCard.

### Full-Text Availability

Electronic versions of many publications, such as *ERIC Digests* and *Research Roundups*, are available free.

Several older books and other publications have been converted into Adobe Acrobat files, which you can download or view directly in a web browser free of charge.

### Acquisitions Information

Authors of material on educational management are invited and shown how to participate in building the ERIC database.